HARCOURT

Math

Practice
Workbook

Grade 1

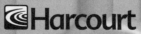

Orlando   Austin   Chicago   New York   Toronto   London   San Diego

Visit *The Learning Site!*
www.harcourtschool.com

CONTENTS

Model Addition Stories

Use ● to show the story.

Draw the ●.

Write the numbers.

1.

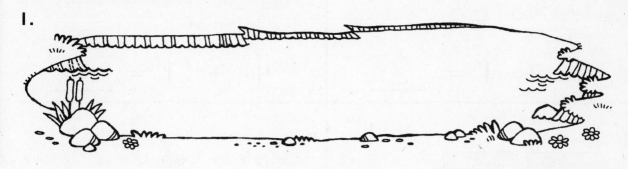

1 big duck   2 little ducks   __3__ in all

2.

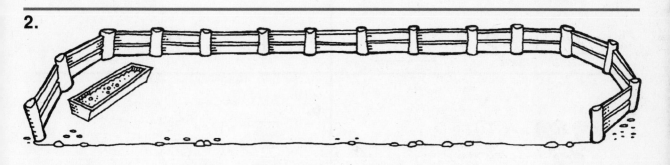

2 pigs   3 pigs come   _____ in all

 Mixed Review

Write the missing numbers.

3. 2, 3, _____

4. 7, 8, _____

5. 3, _____, 5

6. _____, 3, 4

Use Symbols to Add

Add. Write the sum.

1.

1 + 1 = __2__

2.

4 + 1 = ____

3.

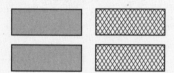

2 + 2 = ____

4.

4 + 2 = ____

5.

3 + 2 = ____

6.

2 + 1 = ____

▶ **Mixed Review**

Write the number that is between.

7. 7, _____, 9   8. 3, _____, 5   9. 6, _____, 8

10. 5, _____, 7   11. 8, _____, 10   12. 4, _____, 6

Algebra: Add 0

Draw circles to show each number.
Write the sum.

1.

$1 + 0 = \underline{\ 1\ }$

2.

$1 + 2 = \underline{\ \ \ }$

3.

$1 + 3 = \underline{\ \ \ }$

4.

$3 + 0 = \underline{\ \ \ }$

5.

$0 + 0 = \underline{\ \ \ }$

6.

$0 + 2 = \underline{\ \ \ }$

7.

$0 + 5 = \underline{\ \ \ }$

8.

$4 + 0 = \underline{\ \ \ }$

9.

$0 + 1 = \underline{\ \ \ }$

10.

$2 + 1 = \underline{\ \ \ }$

11.

$2 + 2 = \underline{\ \ \ }$

12.

$2 + 0 = \underline{\ \ \ }$

▶ **Mixed Review**

Write the number that comes next.

13. 3, 4, 5, ____

14. 5, 4, 3, ____

15. 4, 3, 2, ____

16. 6, 5, 4, ____

Problem Solving • Write a Number Sentence

Draw a picture.
Then write an addition sentence to solve.

1. Liz has 1 big fish.
   She gets 1 little fish.
   How many fish does
   she have in all?

   $\underline{1} \oplus \underline{1} \ominus \underline{2}$

2. 2 yellow ducks swim.
   1 white duck joins them.
   How many ducks are
   there in all?

   ____ ◯ ____ ◯ ____

3. 4 brown snakes lie in the sun.
   1 more snake joins them.
   How many snakes are
   there in all?

   ____ ◯ ____ ◯ ____

4. 3 big cats sleep.
   3 small cats join them.
   How many cats are
   there in all?

   ____ ◯ ____ ◯ ____

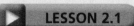

Algebra: Add in Any Order

Use  and ▦ to add.
Circle the addition sentences
that have the same sum.

1. (2 + 1 = _3_)

2. 1 + 0 = _1_

3. (1 + 2 = _3_)

4. 0 + 3 = ____

5. 3 + 1 = ____

6. 1 + 3 = ____

7. 2 + 4 = ____

8. 4 + 1 = ____

9. 4 + 2 = ____

10. 0 + 1 = ____

11. 4 + 0 = ____

12. 0 + 4 = ____

13. 3 + 2 = ____

14. 2 + 3 = ____

15. 2 + 0 = ____

16. 5 + 1 = ____

17. 1 + 5 = ____

18. 3 + 0 = ____

▶ **Mixed Review**

Circle the greater number.

19. 9 or 8   **20.** 4 or 7   **21.** 10 or 11

22. 5 or 7   **23.** 6 or 3   **24.** 12 or 10

Ways to Make 7 and 8

Use 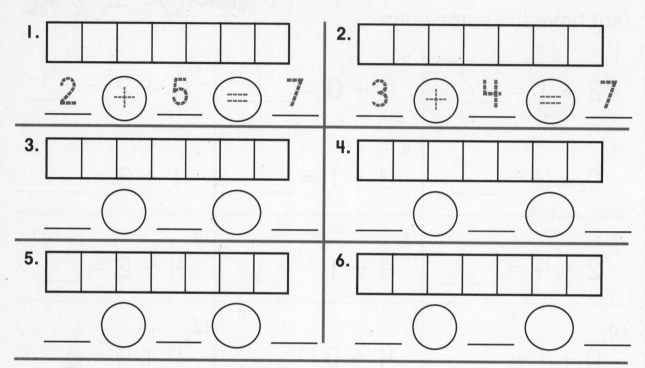 and ⬜ to make 7.
Color. Write the addition sentence.

1.

☐☐☐☐☐☐☐

_2_ ⊕ _5_ ⊜ _7_

2.

☐☐☐☐☐☐☐

_3_ ⊕ _4_ ⊜ _7_

3.

☐☐☐☐☐☐☐

___ ◯ ___ ◯ ___

4.

☐☐☐☐☐☐☐

___ ◯ ___ ◯ ___

5.

☐☐☐☐☐☐☐

___ ◯ ___ ◯ ___

6.

☐☐☐☐☐☐☐

___ ◯ ___ ◯ ___

Use ⬛ and ⬜ to make 8.
Color. Write the addition sentence.

7.

☐☐☐☐☐☐☐☐

___ ◯ ___ ◯ ___

8.

☐☐☐☐☐☐☐☐

___ ◯ ___ ◯ ___

9.

☐☐☐☐☐☐☐☐

___ ◯ ___ ◯ ___

10.

☐☐☐☐☐☐☐☐

___ ◯ ___ ◯ ___

11.

☐☐☐☐☐☐☐☐

___ ◯ ___ ◯ ___

12.

☐☐☐☐☐☐☐☐

___ ◯ ___ ◯ ___

Name _____

Ways to Make 9 and 10

Use Workmat 7, ●, and ○
to make 9 or 10. Draw and color.
Write the addition sentence.

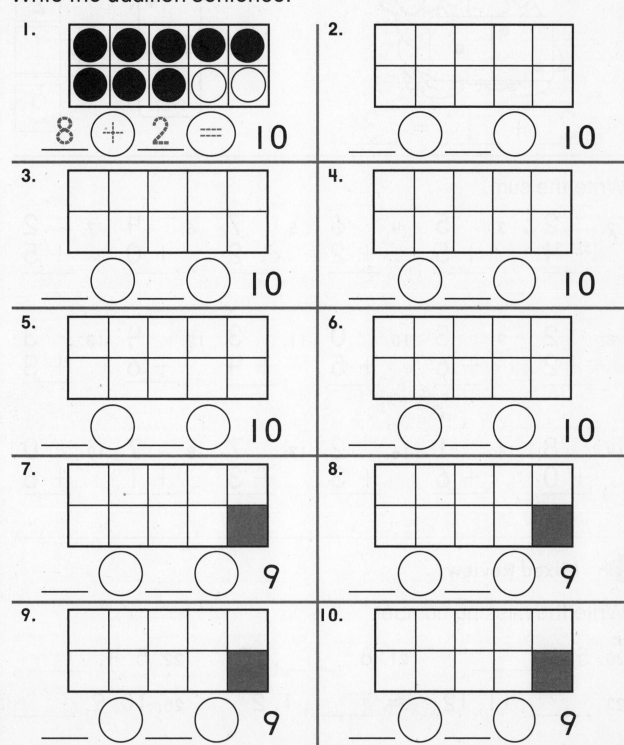

1. __8__ + __2__ = 10

2. ___ ○ ___ ○ 10

3. ___ ○ ___ ○ 10

4. ___ ○ ___ ○ 10

5. ___ ○ ___ ○ 10

6. ___ ○ ___ ○ 10

7. ___ ○ ___ ○ 9

8. ___ ○ ___ ○ 9

9. ___ ○ ___ ○ 9

10. ___ ○ ___ ○ 9

Vertical Addition

Write the numbers to match the dots. Write the sum.

1.

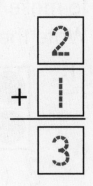

$$\underline{2} + \underline{1} = \underline{3}$$

Write the sum.

2. $\begin{array}{r} 2 \\ +4 \\ \hline \end{array}$
3. $\begin{array}{r} 5 \\ +5 \\ \hline \end{array}$
4. $\begin{array}{r} 6 \\ +2 \\ \hline \end{array}$
5. $\begin{array}{r} 7 \\ +2 \\ \hline \end{array}$
6. $\begin{array}{r} 4 \\ +0 \\ \hline \end{array}$
7. $\begin{array}{r} 2 \\ +5 \\ \hline \end{array}$

8. $\begin{array}{r} 2 \\ +2 \\ \hline \end{array}$
9. $\begin{array}{r} 3 \\ +6 \\ \hline \end{array}$
10. $\begin{array}{r} 0 \\ +5 \\ \hline \end{array}$
11. $\begin{array}{r} 3 \\ +4 \\ \hline \end{array}$
12. $\begin{array}{r} 4 \\ +6 \\ \hline \end{array}$
13. $\begin{array}{r} 3 \\ +3 \\ \hline \end{array}$

14. $\begin{array}{r} 8 \\ +0 \\ \hline \end{array}$
15. $\begin{array}{r} 1 \\ +6 \\ \hline \end{array}$
16. $\begin{array}{r} 2 \\ +3 \\ \hline \end{array}$
17. $\begin{array}{r} 7 \\ +3 \\ \hline \end{array}$
18. $\begin{array}{r} 7 \\ +1 \\ \hline \end{array}$
19. $\begin{array}{r} 0 \\ +6 \\ \hline \end{array}$

▶ **Mixed Review**

Write the missing number.

20. 5, 6, _____

21. 8, _____, 10

22. 5, 4, _____

23. _____, 11, 12

24. _____, 1, 2

25. 10, 9, _____

Problem Solving • Make a Model

Use to show each price.
Draw how many ⓘ¢ there are in all.

1. How much will you spend?

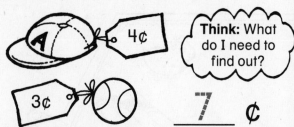

Think: What do I need to find out?

7 ¢

ⓘ¢ ⓘ¢ ⓘ¢

ⓘ¢ ⓘ¢ ⓘ¢ ⓘ¢

2. How much do these cost altogether?

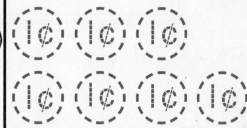

_____ ¢

3. How much will you spend for both?

_____ ¢

▶ **Mixed Review**

Write the sum.

4. $\begin{array}{r} 5 \\ +3 \\ \hline \end{array}$ 5. $\begin{array}{r} 3 \\ +2 \\ \hline \end{array}$ 6. $\begin{array}{r} 0 \\ +7 \\ \hline \end{array}$ 7. $\begin{array}{r} 1 \\ +8 \\ \hline \end{array}$

Name _____

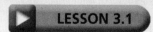

Model Subtraction Stories

Use ● to show a story. Draw the ●.
Cross out how many go away.
Write how many are left.

1.

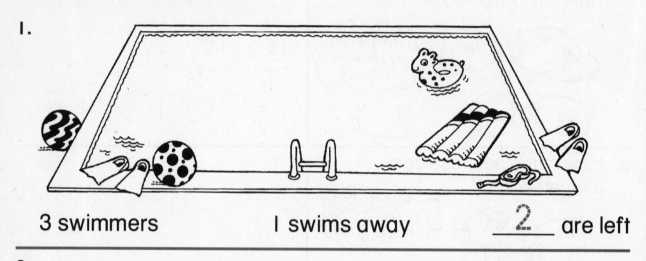

3 swimmers   I swims away   __2__ are left

2.

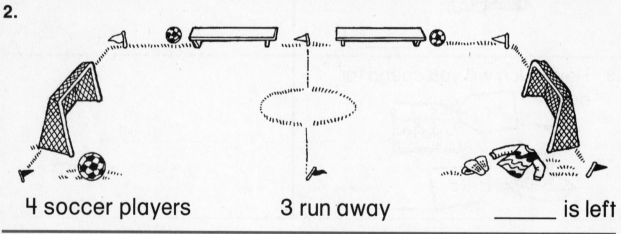

4 soccer players   3 run away   _____ is left

3.

5 skaters   I skates away   _____ are left

Use Symbols to Subtract

Cross out pictures to subtract. Write the difference.

1.

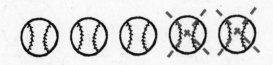

5 – 2 = __3__

2.

2 – 1 = ____

3.

3 – 1 = ____

4.

5 – 4 = ____

5.

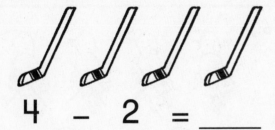

4 – 2 = ____

6.

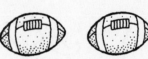

3 – 2 = ____

7.

6 – 5 = ____

8.

4 – 1 = ____

▶ **Mixed Review**

Circle the number that is greater.

9. 5 or 3   10. 5 or 6   11. 4 or 8

12. 8 or 5   13. 7 or 9   14. 2 or 4

Algebra • Write Subtraction Sentences

Write the subtraction sentence.

1.

4 ___ ⃝ 2 ___ ⃝ 2 ___

2.

___ ⃝ ___ ⃝ ___

3.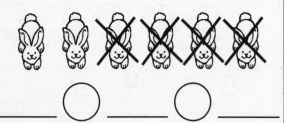

___ ⃝ ___ ⃝ ___

4.

___ ⃝ ___ ⃝ ___

5.

___ ⃝ ___ ⃝ ___

6.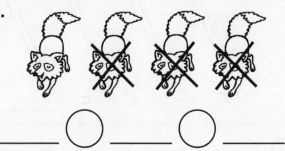

___ ⃝ ___ ⃝ ___

▶ **Mixed Review**

Circle the number that is less.

7. ⑤ or 6　　　**8.** 9 or 7　　　**9.** 2 or 3

Circle the number that is greater.

10. 5 or ⑥　　　**11.** 8 or 9　　　**12.** 5 or 7

Problem Solving • Make a Model

Use ● to subtract. Draw the ●.
Write the difference.

1. Pam has 3 pails.
She gives 1 away.
How many pails does
she have now?

   Think: How can I solve the problem?

   ____2____ pails

2. 2 dogs are on the beach.
1 dog runs away. How
many dogs are left?

   _____ dog

3. There are 6 surfboards.
2 break. How many
surfboards are left?

   _____ surfboards

4. 5 sea gulls are eating.
3 fly away. How many
sea gulls are left?

   _____ sea gulls

Name _____

Algebra • Subtract All or Zero

Write the difference.

1.

$4 - 4 = \underline{0}$

All swim away.

2.

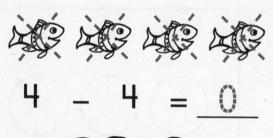

$4 - 0 = \underline{4}$

None swim away.

3.

$3 - 3 = \underline{\hphantom{0}}$

4.

$3 - 0 = \underline{\hphantom{0}}$

5.

$2 - 2 = \underline{\hphantom{0}}$

6.

$2 - 0 = \underline{\hphantom{0}}$

7.

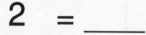

$6 - 0 = \underline{\hphantom{0}}$

8.

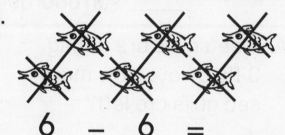

$6 - 6 = \underline{\hphantom{0}}$

Take Apart 7 and 8

Use 🎲 to show all the ways to subtract from 7 and from 8.
Complete the subtraction sentences.

1. $7 - \underline{} = \underline{}$

2. $7 - \underline{} = \underline{}$

3. $7 - \underline{} = \underline{}$

4. $7 - \underline{} = \underline{}$

5. $7 - \underline{} = \underline{}$

6. $7 - \underline{} = \underline{}$

7. $7 - \underline{} = \underline{}$

8. $7 - \underline{} = \underline{}$

9. $8 - \underline{} = \underline{}$

10. $8 - \underline{} = \underline{}$

11. $8 - \underline{} = \underline{}$

12. $8 - \underline{} = \underline{}$

13. $8 - \underline{} = \underline{}$

14. $8 - \underline{} = \underline{}$

15. $8 - \underline{} = \underline{}$

16. $8 - \underline{} = \underline{}$

17. $8 - \underline{} = \underline{}$

▶ **Mixed Review**

Solve.

18. $2 + 2 = \underline{}$

19. $3 + 2 = \underline{}$

20. $6 + 2 = \underline{}$

21. $4 + 3 = \underline{}$

22. $6 + 1 = \underline{}$

23. $4 + 4 = \underline{}$

Take Apart 9 and 10

Use 🎲 to show the ways to subtract from 9 and from 10.
Complete the subtraction sentences.

1. $9 - \underline{0} = \underline{9}$ 2. $9 - \underline{} = \underline{}$

3. $9 - \underline{} = \underline{}$ 4. $9 - \underline{} = \underline{}$

5. $9 - \underline{} = \underline{}$ 6. $9 - \underline{} = \underline{}$

7. $9 - \underline{} = \underline{}$ 8. $9 - \underline{} = \underline{}$

9. $9 - \underline{} = \underline{}$ 10. $9 - \underline{} = \underline{}$

11. $10 - \underline{} = \underline{}$ 12. $10 - \underline{} = \underline{}$

13. $10 - \underline{} = \underline{}$ 14. $10 - \underline{} = \underline{}$

15. $10 - \underline{} = \underline{}$ 16. $10 - \underline{} = \underline{}$

17. $10 - \underline{} = \underline{}$ 18. $10 - \underline{} = \underline{}$

▶ **Mixed Review**

Solve.

19. $7 - 2 = \underline{}$ 20. $8 - 1 = \underline{}$ 21. $3 - 1 = \underline{}$

22. $5 - 2 = \underline{}$ 23. $1 - 1 = \underline{}$ 24. $6 - 4 = \underline{}$

Vertical Subtraction

Write the difference.

1. $\begin{array}{r} 7 \\ -2 \\ \hline \end{array}$
2. $\begin{array}{r} 8 \\ -1 \\ \hline \end{array}$
3. $\begin{array}{r} 5 \\ -4 \\ \hline \end{array}$
4. $\begin{array}{r} 9 \\ -6 \\ \hline \end{array}$
5. $\begin{array}{r} 6 \\ -6 \\ \hline \end{array}$

6. $\begin{array}{r} 4 \\ -2 \\ \hline \end{array}$
7. $\begin{array}{r} 10 \\ -3 \\ \hline \end{array}$
8. $\begin{array}{r} 8 \\ -0 \\ \hline \end{array}$
9. $\begin{array}{r} 3 \\ -2 \\ \hline \end{array}$
10. $\begin{array}{r} 6 \\ -3 \\ \hline \end{array}$

11. $\begin{array}{r} 10 \\ -5 \\ \hline \end{array}$
12. $\begin{array}{r} 9 \\ -2 \\ \hline \end{array}$
13. $\begin{array}{r} 7 \\ -4 \\ \hline \end{array}$
14. $\begin{array}{r} 10 \\ -4 \\ \hline \end{array}$
15. $\begin{array}{r} 8 \\ -5 \\ \hline \end{array}$

16. $\begin{array}{r} 9 \\ -5 \\ \hline \end{array}$
17. $\begin{array}{r} 5 \\ -2 \\ \hline \end{array}$
18. $\begin{array}{r} 7 \\ -6 \\ \hline \end{array}$
19. $\begin{array}{r} 4 \\ -0 \\ \hline \end{array}$
20. $\begin{array}{r} 3 \\ -3 \\ \hline \end{array}$

▶ **Mixed Review**

Solve.

21. $1 + 1 =$ _____
22. $3 + 2 =$ _____
23. $4 + 3 =$ _____

24. $4 + 5 =$ _____
25. $3 + 3 =$ _____
26. $2 + 1 =$ _____

Name _____

 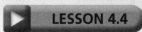

Subtract to Compare

Draw lines to match.
Subtract to find how many more.

1.

$7 - 5 = \underline{2}$

$\underline{2}$ more

2.

$6 - 2 = \underline{\hspace{1cm}}$

_____ more

3.

$5 - 4 = \underline{\hspace{1cm}}$

_____ more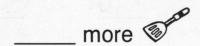

4.

$4 - 1 = \underline{\hspace{1cm}}$

_____ more

▶ **Mixed Review**

Solve.

5. $7 + 3 = \underline{\hspace{1cm}}$ 6. $8 + 2 = \underline{\hspace{1cm}}$ 7. $5 + 5 = \underline{\hspace{1cm}}$

8. $6 + 2 = \underline{\hspace{1cm}}$ 9. $2 + 5 = \underline{\hspace{1cm}}$ 10. $4 + 1 = \underline{\hspace{1cm}}$

Problem Solving • Draw a Picture

Draw a picture to solve the problem.
Write how many were given away.

1. Pedro has 5 paper clips.
He gives some to his friend.
He has 1 paper clip left. How
many paper clips does Pedro
give to his friend?

____4____ paper clips

2. Molly has 8 pencils. She gives
some to her friends. She has
4 pencils left. How many pencils
does Molly give to her friends?

_____ pencils

3. Tina has 10 pieces of chalk. She
gives some to her friends. She has
2 pieces left. How many pieces of
chalk does Tina give to her friends?

_____ pieces of chalk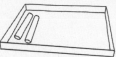

4. There are 7 books on the shelf.
Mike takes some books to read.
There are 5 books left. How many
books does Mike take to read?

_____ books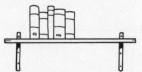

Count On 1 and 2

Use ●. Count on. Find the sum.

1. $7 + 2 = \underline{9}$ | 2. $6 + 1 = \underline{7}$

3. $4 + 2 = \underline{}$ 4. $5 + 1 = \underline{}$ 5. $8 + 2 = \underline{}$

6. $2 + 2 = \underline{}$ 7. $1 + 2 = \underline{}$ 8. $8 + 1 = \underline{}$

9. $5 + 2 = \underline{}$ 10. $3 + 1 = \underline{}$ 11. $7 + 1 = \underline{}$

12. $6 + 2 = \underline{}$ 13. $4 + 1 = \underline{}$ 14. $3 + 2 = \underline{}$

▶ **Mixed Review**

Write the number that is one more.

15. $6,$ _____   16. $4,$ _____   17. $2,$ _____

18. $5,$ _____   19. $7,$ _____   20. $9,$ _____

Use a Number Line to Count On

Use the number line.
Count on to find the sum.

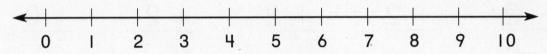

0   1   2   3   4   5   6   7   8   9   10

1. 6
   +3

2. 3
   +2

3. 5
   +2

4. 7
   +1

5. 4
   +2

6. 7
   +3

7. 6
   +2

8. 8
   +2

9. 1
   +2

10. 4
   +1

11. 4
   +3

12. 8
   +1

13. 7
   +2

14. 5
   +3

15. 3
   +1

▶ **Mixed Review**

Write the missing number.

16. 2, 3, ____, 5

17. 7, 8, 9, ____

18. ____, 5, 6, 7

19. 3, 4, ____, 6, 7

Use Doubles

Add. Then circle the doubles facts.

1. 2
    + 2
    ———
     4

2. 3
    + 2

3. 8
    + 2

4. 4
    + 2

5. 0
    + 0

6. 9
    + 1

7. 7
    + 3

8. 1
    + 9

9. 2
    + 6

10. 5
    + 5

11. 5
    + 2

12. 1
    + 1

13. 7
    + 2

14. 4
    + 1

15. 4
    + 3

16. 5
    + 3

17. 8
    + 1

18. 3
    + 3

19. 1
    + 6

20. 4
    + 4

▶ **Mixed Review**

Solve.

21. $3 + 2 =$ _____

22. $5 + 2 =$ _____

23. $6 + 1 =$ _____

24. $4 - 3 =$ _____

25. $7 - 2 =$ _____

26. $4 - 2 =$ _____

Problem Solving • Draw a Picture

Draw a picture to solve.
Write an addition sentence to check.

1. 2 seagulls are on the beach.
   3 more join them.
   How many seagulls are there?

   __2__ ⊕ __3__ ⊜ __5__
   seagulls

   > **Think**
   > What numbers will help me solve the problem?

2. There are 3 fish swimming.
   4 more swim with them.
   How many fish are there?

   ____ ◯ ____ ◯ ____
   fish

3. There are 2 turtles on the sand.
   6 more join them.
   How many turtles are there?

   ____ ◯ ____ ◯ ____
   turtles

4. There are 4 big beach balls.
   There are 5 little beach balls.
   How many beach balls are there?

   ____ ◯ ____ ◯ ____
   beach balls

Use the Strategies

Add. Write the sums.

1. 4
   + 3

    7

2. 3
   + 2

3. 3
   + 3

4. 4
   + 1

5. 4
   + 2

6. 5
   + 3

7. 6
   + 4

8. 7
   + 2

9. 9
   + 1

10. 8
    + 0

11. 5
    + 5

12. 3
    + 4

13. 3
    + 1

14. 2
    + 2

15. 5
    + 2

16. 1
    + 1

17. 4
    + 4

18. 1
    + 2

19. 5
    + 1

20. 6
    + 2

► **Mixed Review**

Solve.

21. $5 + 5 =$ ___

22. $4 + 5 =$ ___

23. $2 + 6 =$ ___

24. $4 + 2 =$ ___

25. $7 + 1 =$ ___

26. $3 + 3 =$ ___

27. $1 + 5 =$ ___

28. $3 + 5 =$ ___

29. $4 + 4 =$ ___

Sums to 8

Add. Use the key.

Color each balloon by the sum.

What patterns do you see?

5	red
6	blue
7	red
8	blue

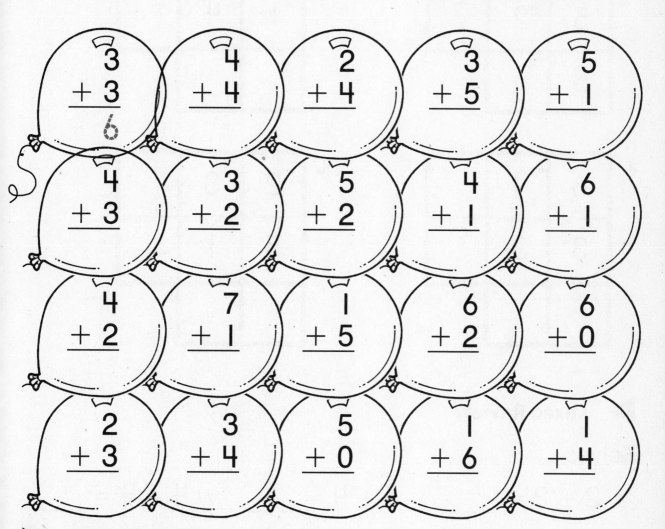

Row 1: 3 + 3 = 6 | 4 + 4 | 2 + 4 | 3 + 5 | 5 + 1

Row 2: 4 + 3 | 3 + 2 | 5 + 2 | 4 + 1 | 6 + 1

Row 3: 4 + 2 | 7 + 1 | 1 + 5 | 6 + 2 | 6 + 0

Row 4: 2 + 3 | 3 + 4 | 5 + 0 | 1 + 6 | 1 + 4

► **Mixed Review**

Solve.

1. $2 + 2 =$ ___   2. $2 + 1 =$ ___   3. $3 + 3 =$ ___

4. $7 + 0 =$ ___   5. $1 + 1 =$ ___   6. $3 + 1 =$ ___

Sums to 10

Add across. Add down. Write the sums.

1.

3	4	7
5	2	7
8	6	

2.

8	2	
1	4	

3.

6	4	
3	3	

4.

2	3	
5	1	

▶ **Mixed Review**

Solve.

5. $2 + 2 =$ _____   6. $5 + 4 =$ _____   7. $4 + 4 =$ _____

8. $6 + 2 =$ _____   9. $3 + 4 =$ _____   10. $4 + 5 =$ _____

11. $1 + 6 =$ _____   12. $0 + 5 =$ _____   13. $7 + 3 =$ _____

Algebra • Follow the Rule

Complete the table.
Follow the rule.

1.

Add 1	
6	7
7	8
8	9

2.

Add 6	
1	
2	
3	

3.

Add 5	
2	
3	
4	

4.

Add 2	
2	
4	
6	

5.

Add 4	
1	
3	
5	

6.

Add 3	
3	
4	
5	

▶ **Mixed Review**

Solve.

7. $3 + 2 =$ _____

8. $4 + 4 =$ _____

9. $2 + 1 =$ _____

10. $1 + 8 =$ _____

11. $6 + 3 =$ _____

12. $2 + 2 =$ _____

13. $1 + 1 =$ _____

14. $5 + 2 =$ _____

15. $4 + 5 =$ _____

Problem Solving • Write a Number Sentence

Solve. Write a number sentence.
Draw a picture to check.

Think
What do I need to solve the problem?

1. There are 5 basketballs.
   Laura sees 1 more.
   How many basketballs
   are there in all?

   ____ ◯ ____ ◯ ____
   basketballs

2. There are 7 swings.
   2 more are added.
   How many swings
   are there now?

   ____ ◯ ____ ◯ ____ swings

3. 4 birds are in a tree.
   3 more birds join them.
   How many birds are
   there now?

   ____ ◯ ____ ◯ ____ birds

4. 4 children sit in a circle.
   4 more join them.
   How many children
   are in the circle now?

   ____ ◯ ____ ◯ ____ children

Name _____

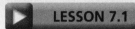

Use a Number Line to Count Back 1 and 2

Use the number line.
Count back to subtract.

1.

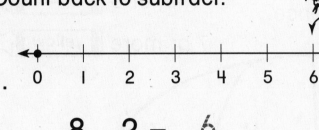

$$8 - 2 = \underline{6}$$

Start at 8. Count back 2.
Where are you?

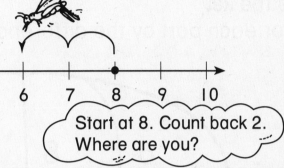

2.

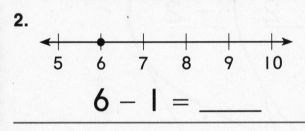

$$6 - 1 = \underline{}$$

3.

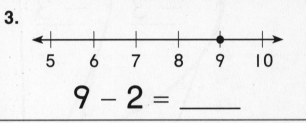

$$9 - 2 = \underline{}$$

4.

$$2 - 2 = \underline{}$$

5.

$$5 - 2 = \underline{}$$

6.

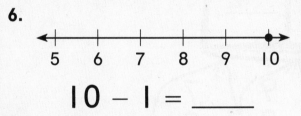

$$10 - 1 = \underline{}$$

7.

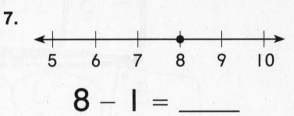

$$8 - 1 = \underline{}$$

▶ **Mixed Review**

Solve.

8. $2 + 3 = \underline{}$ **9.** $4 + 2 = \underline{}$ **10.** $2 + 2 = \underline{}$

11. $6 + 3 = \underline{}$ **12.** $5 + 5 = \underline{}$ **13.** $7 + 1 = \underline{}$

14. $2 + 1 = \underline{}$ **15.** $6 + 2 = \underline{}$ **16.** $5 + 3 = \underline{}$

Use a Number Line to Count Back 3

Count back to subtract.
Use the key.
Color each part by the difference.

3 or less | red
4, 5, or 6 | blue
7 or more | yellow

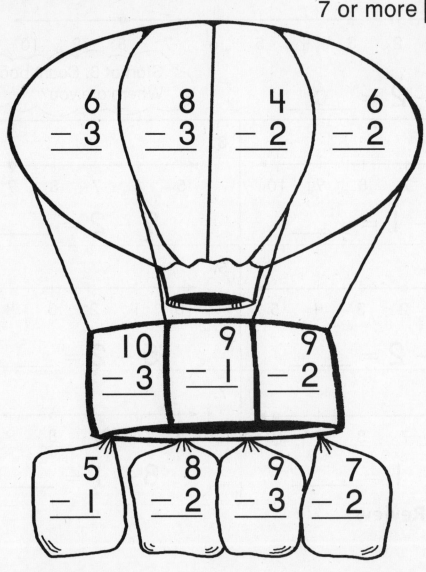

▶ **Mixed Review**

Write the missing number.

1. 1, 3, 5, 7, _____   2. 1, 1, 2, 2, 3, 3, 4, _____

3. 2, 4, 6, 8, _____   4. 1, 2, 2, 3, 4, 4, 5, 6, _____

PW30 **Practice**

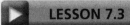

Algebra • Relate Addition and Subtraction

Add. Then subtract.

1.

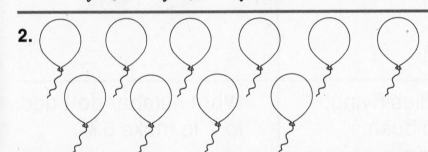

$5 + 3 =$ __8__

$8 - 3 =$ __5__

2.

$6 + 4 =$ ____

$10 - 4 =$ ____

3.

$5 + 4 =$ ____

$9 - 4 =$ ____

4. $\begin{array}{r} 7 \\ +1 \\ \hline \end{array}$ $\begin{array}{r} 8 \\ -1 \\ \hline \end{array}$
5. $\begin{array}{r} 3 \\ +3 \\ \hline \end{array}$ $\begin{array}{r} 6 \\ -3 \\ \hline \end{array}$
6. $\begin{array}{r} 4 \\ +3 \\ \hline \end{array}$ $\begin{array}{r} 7 \\ -3 \\ \hline \end{array}$

▶ **Mixed Review**

Write the numbers before and after.

7. ____, 4, ____

8. ____, 22, ____

9. ____, 47, ____

10. ____, 11, ____

11. ____, 35, ____

12. ____, 59, ____

Problem Solving • Draw a Picture

Draw a picture to solve the problem.

1. There were 8 seagulls flying.
   Some landed on the sand.
   3 seagulls are still flying.
   How many seagulls landed
   on the sand?

   __5__ seagulls

What number do I add
to 3 to make 8?

---

2. There were 5 fireflies flying.
   Some landed on a bush.
   1 firefly is still flying.
   How many fireflies landed
   on a bush?

   _____ fireflies

What number do I add
to 1 to make 5?

---

3. There were 9 balloons floating.
   Some balloons popped.
   6 balloons are still in the air.
   How many balloons
   popped?

   _____ balloons

What number do I add
to 6 to make 9?

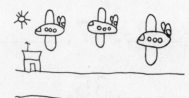

---

4. There were 7 airplanes flying.
   Some landed at the airport.
   3 airplanes are still flying.
   How many landed?

   _____ airplanes

What number do I add
to 3 to make 7?

Name _____

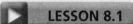

Use the Strategies

Circle the facts for **subtract 0** and for **subtract all**. Subtract. Write the difference.

1.

$$\begin{array}{r} 5 \\ -\ 0 \\ \hline 5 \end{array}$$

$$\begin{array}{r} 6 \\ -\ 3 \\ \hline \end{array}$$

$$\begin{array}{r} 7 \\ -\ 3 \\ \hline \end{array}$$

$$\begin{array}{r} 8 \\ -\ 2 \\ \hline \end{array}$$

$$\begin{array}{r} 9 \\ -\ 3 \\ \hline \end{array}$$

$$\begin{array}{r} 10 \\ -10 \\ \hline \end{array}$$

2.

$$\begin{array}{r} 5 \\ -\ 2 \\ \hline \end{array}$$

$$\begin{array}{r} 9 \\ -\ 0 \\ \hline \end{array}$$

$$\begin{array}{r} 7 \\ -\ 2 \\ \hline \end{array}$$

$$\begin{array}{r} 7 \\ -\ 7 \\ \hline \end{array}$$

$$\begin{array}{r} 9 \\ -\ 2 \\ \hline \end{array}$$

$$\begin{array}{r} 5 \\ -\ 1 \\ \hline \end{array}$$

3.

$$\begin{array}{r} 10 \\ -\ 1 \\ \hline \end{array}$$

$$\begin{array}{r} 6 \\ -\ 2 \\ \hline \end{array}$$

$$\begin{array}{r} 8 \\ -\ 8 \\ \hline \end{array}$$

$$\begin{array}{r} 8 \\ -\ 3 \\ \hline \end{array}$$

$$\begin{array}{r} 6 \\ -\ 0 \\ \hline \end{array}$$

$$\begin{array}{r} 10 \\ -\ 3 \\ \hline \end{array}$$

4.

$$\begin{array}{r} 10 \\ -\ 0 \\ \hline \end{array}$$

$$\begin{array}{r} 9 \\ -\ 1 \\ \hline \end{array}$$

$$\begin{array}{r} 8 \\ -\ 1 \\ \hline \end{array}$$

$$\begin{array}{r} 7 \\ -\ 1 \\ \hline \end{array}$$

$$\begin{array}{r} 6 \\ -\ 0 \\ \hline \end{array}$$

$$\begin{array}{r} 10 \\ -\ 2 \\ \hline \end{array}$$

▶ **Mixed Review**

Solve.

5. $3 + 2 =$ _____ $4 - 3 =$ _____ $2 + 5 =$ _____

6. $3 + 5 =$ _____ $4 + 4 =$ _____ $4 + 2 =$ _____

Subtraction to 10

Subtract across. Subtract down.

1.

6	2	4
5	2	3
1	0	1

2.

10	5	
4	2	

3.

8	3	
6	2	

4.

9	2	
5	0	

▶ **Mixed Review**

Put a + or − in the circle to make
the number sentence correct.

5. 5 ◯ 3 = 8   6 ◯ 2 = 4   7 ◯ 3 = 4

6. 10 ◯ 2 = 8   4 ◯ 5 = 9   7 ◯ 2 = 5

7. 8 ◯ 2 = 10   4 ◯ 3 = 1   4 ◯ 4 = 8

Algebra • Follow the Rule

Complete the table. Follow the rule.

1.

Subtract 2	
8	6
9	7
10	

2.

Subtract 5	
10	
8	
6	

3.

Subtract 4	
9	
7	
5	

4.

Subtract 3	
7	
6	
5	

5.

Subtract 0	
8	
6	
4	

6.

Subtract 1	
6	
8	
10	

▶ **Mixed Review**

Solve.

7. $7 + 2 =$ _____ $3 + 5 =$ _____ $2 + 5 =$ _____

8. $5 - 3 =$ _____ $7 - 5 =$ _____ $9 - 1 =$ _____

9. $6 - 3 =$ _____ $8 - 5 =$ _____ $4 - 2 =$ _____

10. $5 + 5 =$ _____ $6 + 2 =$ _____ $1 + 4 =$ _____

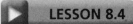

Fact Families to 10

Add or subtract.

Write the numbers in the **fact family**.

1.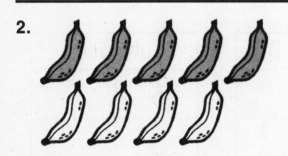

$$\begin{array}{r} 4 \\ + 2 \\ \hline 6 \end{array} \quad \begin{array}{r} 2 \\ + 4 \\ \hline 6 \end{array} \quad \begin{array}{r} 6 \\ - 2 \\ \hline 4 \end{array} \quad \begin{array}{r} 6 \\ - 4 \\ \hline 2 \end{array}$$

| 4 | 2 | 6 |

2.

$$\begin{array}{r} 5 \\ + 4 \\ \hline \end{array} \quad \begin{array}{r} 4 \\ + 5 \\ \hline \end{array} \quad \begin{array}{r} 9 \\ - 4 \\ \hline \end{array} \quad \begin{array}{r} 9 \\ - 5 \\ \hline \end{array}$$

| | | |

3.

$$\begin{array}{r} 5 \\ + 3 \\ \hline \end{array} \quad \begin{array}{r} 3 \\ + 5 \\ \hline \end{array} \quad \begin{array}{r} 8 \\ - 3 \\ \hline \end{array} \quad \begin{array}{r} 8 \\ - 5 \\ \hline \end{array}$$

| | | |

▶ **Mixed Review**

Solve.

4. $1 + 7 =$ _____ $2 + 3 =$ _____ $3 + 4 =$ _____

5. $5 - 4 =$ _____ $8 - 3 =$ _____ $9 - 7 =$ _____

Problem Solving • Choose the Operation

Circle **add** or **subtract**.
Write the number sentence to solve.

1. There are 7 cherries.
   Clyde eats 4.
   How many are left?

   __3__ cherries

   add   (subtract)

   $7 \bigcirc 4 \bigcirc 3$
   (− =)

2. There are 4 plates at the
   table. Leah adds 2 more.
   How many are there now?

   _____ plates

   add   subtract

   ___ ○ ___ ○ ___

3. Andy washes 5 pears.
   Anna washes 3 pears.
   How many pears do they
   wash in all?

   _____ pears

   add   subtract

   ___ ○ ___ ○ ___

4. There are 10 marbles.
   4 roll under the table.
   How many are left?

   _____ marbles

   add   subtract

   ___ ○ ___ ○ ___

Algebra • Sort and Classify

Draw a line from each shape to the group
where it belongs.

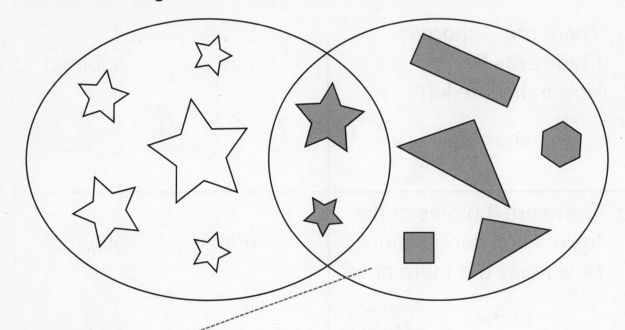

1. 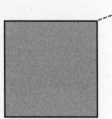 **2.**  **3.**  **4.**

▶ **Mixed Review**

Solve.

5. $3 + 5 =$ _____ $6 - 0 =$ _____ $9 - 2 =$ _____

6. $7 + 0 =$ _____ $4 - 4 =$ _____ $2 + 3 =$ _____

7. $6 - 2 =$ _____ $7 + 2 =$ _____ $5 + 1 =$ _____

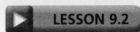

Make Concrete Graphs

Use the graph to answer the questions.

Hats We Like					
🎩	🎩	🎩			
🧢	🧢	🧢	🧢	🧢	🧢
🤠	🤠	🤠	🤠	🤠	

1. How many children chose ? _____ 5

2. How many children chose ? _____

3. Which hat did the most children choose? _____

4. Which hat did the fewest children choose? _____

5. How many children in all chose 🎩 or 🧢? _____

6. How many more children chose 🧢 than 🤠? _____

▶ **Mixed Review**

Solve.

7. $\begin{array}{r} 5 \\ +1 \\ \hline \end{array}$ $\begin{array}{r} 6 \\ -1 \\ \hline \end{array}$

8. $\begin{array}{r} 4 \\ +4 \\ \hline \end{array}$ $\begin{array}{r} 8 \\ -4 \\ \hline \end{array}$

9. $\begin{array}{r} 6 \\ +3 \\ \hline \end{array}$ $\begin{array}{r} 9 \\ -3 \\ \hline \end{array}$

Make Picture Graphs

Toys We Like					
bears 🧸	☺	☺	☺	☺	☺
cars 🚗	☺	☺			
dolls 🪆	☺	☺	☺	☺	

Use the picture graph to answer the questions.

1. How many children chose bears? _____5_____ children

2. How many children chose cars? _____ children

3. How many more children chose
   bears than cars? _____ children

4. Did more children choose

   dolls or cars? _____

▶ **Mixed Review**

Solve.

5. 8 + 2 = _____   3 + 4 = _____   2 + 7 = _____

6. 10 − 4 = _____   9 − 3 = _____   8 − 5 = _____

7. 8 − 0 = _____   9 − 2 = _____   6 − 1 = _____

Read a Tally Table

Complete the tally table.

Kinds of Toys		Total
cars		
trucks		

Each **|** stands for 1 toy.

|||| stands for 5 toys.

1. How many are cars? _____

2. How many are trucks? _____

3. Sort the toys another way. Complete the tally table.

Sizes of Toys		Total

4. Write a question about the Sizes of Toys tally
   table. Have a classmate answer your question.

_____

_____

_____

_____

Make Bar Graphs

1. Write how many tally marks.

Sports We Like		Total
soccer	卌 ||	7
softball	卌	
football	|||	

2. Color the bar graph to match the tally marks.

Sports We Like							
soccer	███	███	███	███	███	███	███
softball							
football							
	0	1	2	3	4	5	6	7

Use the graph to answer
the questions.

_____
_ _ _ _ _ _ _ _ _ _ _ _ _

3. Which sport got 5 choices?

_____

4. Which sport did the most
children choose?

_ _ _ _ _ _ _ _ _ _ _ _ _
_____

5. Which sport did the fewest
children choose?

_ _ _ _ _ _ _ _ _ _ _ _ _
_____

▶ **Mixed Review**

Circle the greater number.

6. 6 or 7   8 or 2   9 or 10

Circle the number that is less.

7. 4 or 2   7 or 9   8 or 5

Problem Solving • Use Data from a Graph

Use the bar graph to answer the questions.

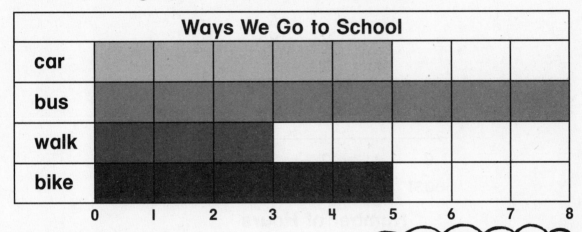

Ways We Go to School

	0	1	2	3	4	5	6	7	8
car									
bus									
walk									
bike									

1. How many children in all go by bike or car?

   Think: I can add to solve a problem.

   ___10___ children

   _5_ ⊕ _5_ ⊜ _10_

2. How many more children go by car than walk?

   Think: I can subtract to solve a problem.

   _____ more children

   ___ ◯ ___ ◯ ___

3. Do more children go by bike or by bus?

   Think: How many go each way?

   _____

   - - - - - - - - - - - - - - - - - -

   _____ children go by bus.

   _____

   _____ children go by bike.

   How many more children go that way?

   Think: I can subtract.

   ___ ◯ ___ ◯ ___

   _____ more children

Interpret Graphs

Use the graph to answer the questions.

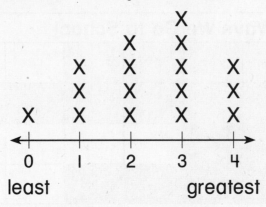

```
                          X
                   X      X
            X      X      X      X
            X      X      X      X
     X      X      X      X      X
    ←+──────+──────+──────+──────+→
     0      1      2      3      4
   least                 greatest
```

**Number of Hours
Reading Books Each Week**

---

1. How many hours do the most children read books? _3_

2. What is the greatest number of hours children read? ____

3. What is the least number of hours children read? ____

4. What is the difference between the greatest number of hours and the least number of hours?

   ____ ◯ ____ ◯ ____

 Mixed Review

Solve.

5. $9 - 4 =$ ____ $9 - 5 =$ ____ $4 + 5 =$ ____

6. $8 - 5 =$ ____ $8 - 3 =$ ____ $5 + 3 =$ ____

7. $10 - 2 =$ ____ $10 - 8 =$ ____ $8 + 2 =$ ____

Name _____

Teen Numbers

Draw the tens and ones.
Write how many tens and ones.

1.

fifteen

____ = ____ ten ____ ones

2.

twelve

____ = ____ ten ____ ones

3.

fourteen

____ = ____ ten ____ ones

4.

seventeen

____ = ____ ten ____ ones

5.

eighteen

____ = ____ ten ____ ones

6.

thirteen

____ = ____ ten ____ ones

▶ **Mixed Review**

Solve.

7. $6 - 4 =$ ____ $8 + 2 =$ ____ $7 - 4 =$ ____

8. $9 - 4 =$ ____ $8 + 1 =$ ____ $6 + 3 =$ ____

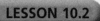

Tens

Count by tens. Write the number.

1.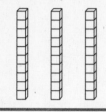
$$\frac{3}{\text{tens}} = \frac{30}{\text{thirty}}$$

2.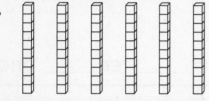
$$\frac{}{\text{tens}} = \frac{}{\text{sixty}}$$

3.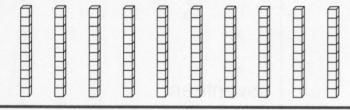
$$\frac{}{\text{tens}} = \frac{}{\text{one hundred}}$$

4.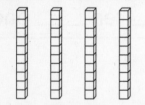
$$\frac{}{\text{tens}} = \frac{}{\text{forty}}$$

5.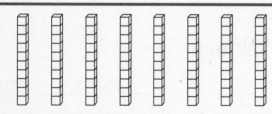
$$\frac{}{\text{tens}} = \frac{}{\text{eighty}}$$

▶ **Mixed Review**

Solve.

6. $3 + 3 =$ _____ $2 + 3 =$ _____ $4 + 4 =$ _____

7. $5 - 3 =$ _____ $7 - 1 =$ _____ $5 - 1 =$ _____

8. $2 + 4 =$ _____ $4 + 1 =$ _____ $6 + 2 =$ _____

Tens and Ones to 50

Write how many tens and ones.
Write the number.

1.

__2__ tens __3__ ones = __23__

2.

____ tens ____ ones = ____

3.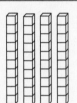

____ tens ____ ones = ____

4.

____ ten ____ ones = ____

5.

____ tens ____ ones = ____

6.

____ tens ____ one = ____

▶ **Mixed Review**

Solve.

7. $3 + 4 =$ _____ $6 + 2 =$ _____ $3 + 3 =$ _____

8. $4 + 4 =$ _____ $5 + 1 =$ _____ $4 + 2 =$ _____

9. $8 - 7 =$ _____ $5 - 2 =$ _____ $6 - 4 =$ _____

10. $8 - 3 =$ _____ $7 - 3 =$ _____ $8 - 1 =$ _____

Tens and Ones to 100

Write how many tens and ones. Write the number.

1.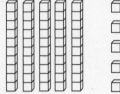

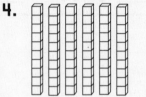 **5** tens **8** ones = **58**

2.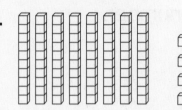

____ tens ____ ones = ____

3.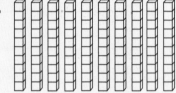

____ tens ____ ones = ____

4.

____ tens ____ ones = ____

5.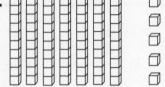

____ tens ____ ones = ____

6.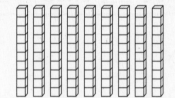

____ tens ____ ones = ____

▶ **Mixed Review**

Solve.

7. $8 + 1 = $ ____ $4 + 5 = $ ____ $3 + 6 = $ ____

8. $2 + 7 = $ ____ $3 + 5 = $ ____ $2 + 6 = $ ____

9. $9 - 1 = $ ____ $9 - 6 = $ ____ $9 - 2 = $ ____

10. $9 - 3 = $ ____ $8 - 3 = $ ____ $9 - 9 = $ ____

Algebra • Different Ways to Make Numbers

Write how many tens and ones.
Write the number in a different way.

1.

__2__ tens __3__ ones = __23__

__20__ + __3__

2.

____ tens ____ ones = ____

____ + ____

3.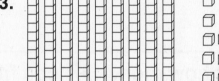

____ tens ____ ones = ____

____ + ____

4.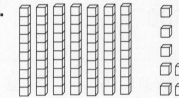

____ tens ____ ones = ____

____ + ____

5.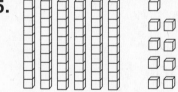

____ tens ____ ones = ____

____ + ____

6.

____ tens ____ one = ____

____ + ____

▶ **Mixed Review**

Write the missing numbers.

7.

8
−☐
___
5

8.
8
− 5
___
☐

9.
5
+☐
___
8

10.
3
+☐
___
8

Problem Solving • Make Reasonable Estimates

Circle the closest estimate.

1. About how many 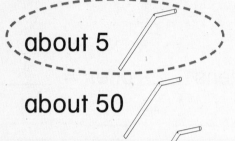 can you hold in one hand?

(about 5)

about 50

about 500

2. About how many can fit in a cup?

about 2

about 20

about 200

3. About how many can you hold in two hands?

about 3

about 30

about 300

4. About how many can fit on a desk?

about 1

about 10

about 100

5. About how many can fit in a cup?

about 3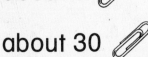

about 30

about 300

6. About how many would ride in one car?

about 5

about 50

about 500

Algebra • Greater Than

Circle the greater number.
Write the numbers.
You can use ▭ ◻ to help.

1. 37 ⟮73⟯

__73__ is greater than __37__.

__73__ > __37__

2. 93 36

_____ is greater than _____.

_____ > _____

3. 16 60

_____ is greater than _____.

_____ > _____

4. 56 59

_____ is greater than _____.

_____ > _____

5. 45 35

_____ is greater than _____.

_____ > _____

6. 19 91

_____ is greater than _____.

_____ > _____

▶ **Mixed Review**

Solve.

7. $4 + 6 =$ _____ $8 - 2 =$ _____ $3 + 5 =$ _____

8. $5 + 5 =$ _____ $4 + 5 =$ _____ $2 + 8 =$ _____

9. $9 - 2 =$ _____ $9 - 1 =$ _____ $4 + 4 =$ _____

10. $6 + 3 =$ _____ $2 + 7 =$ _____ $7 + 3 =$ _____

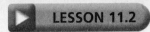

Algebra • Less Than

Circle the number that is less.
Write the numbers.
You can use ⬛⬛⬛⬛⬛⬛⬛⬛⬛⬛ ⬜ to help.

1. ⟨19⟩   61

___19___ is less than __61__ .

_19_ < _61_

2. 41   15

_____ is less than _____ .

____ < ____

3. 65   56

_____ is less than _____ .

____ < ____

4. 45   44

_____ is less than _____ .

____ < ____

5. 13   31

_____ is less than _____ .

____ < ____

6. 42   44

_____ is less than _____ .

____ < ____

▶ **Mixed Review**

Solve.

7. $10 - 8 =$ _____ $10 - 7 =$ _____ $10 - 9 =$ _____

8. $10 - 4 =$ _____ $10 - 5 =$ _____ $10 - 6 =$ _____

9. $9 - 3 =$ _____ $9 - 4 =$ _____ $9 - 5 =$ _____

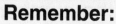

Algebra • Use Symbols to Compare

Compare the numbers.
Use 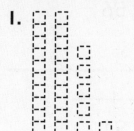 to show each
number. Draw the .
Write <, >, or = in the circle.

Remember:
< means **is less than**.
> means **is greater than**.
= means **is equal to**.

1.

26 (>) 19

2.

12 ◯ 32

3.

24 ◯ 42

4.

17 ◯ 17

5.

83 ◯ 71

6.

23 ◯ 37

▶ **Mixed Review**

Solve.

7. $7 - 3 = $ _____ $10 - 4 = $ _____ $9 - 3 = $ _____

8. $10 - 6 = $ _____ $9 - 7 = $ _____ $8 - 3 = $ _____

Order on a Number Line

Write the number that is just before,
between, or just after.

1.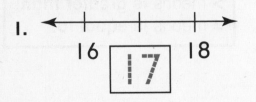
16 [17] 18

2.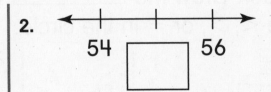
54 [ ] 56

3.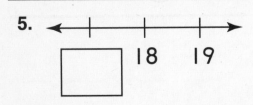
75 [ ] 77

4.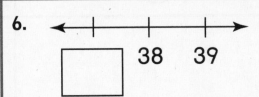
89 [ ] 91

5.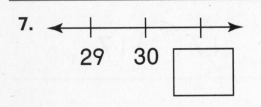
[ ] 18   19

6.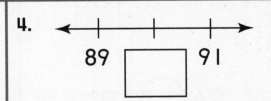
[ ]   38   39

7.
29   30   [ ]

8.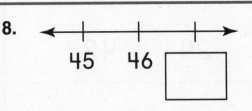
45   46   [ ]

▶ **Mixed Review**

Solve.

9. $5 + 1 =$ _____ $8 + 2 =$ _____ $7 + 3 =$ _____

10. $10 - 3 =$ _____ $10 - 2 =$ _____ $10 - 1 =$ _____

11. $3 + 3 =$ _____ $4 + 4 =$ _____ $5 + 5 =$ _____

Count Forward and Backward

Write the numbers.
Count forward.

1. 32, _33_, ___, ___, ___, ___, ___, ___, ___, ___

2. 21, ___, ___, ___, ___, ___, ___, ___, ___, ___

3. 45, ___, ___, ___, ___, ___, ___, ___, ___, ___

4. 88, ___, ___, ___, ___, ___, ___, ___, ___, ___

Count backward.

5. 71, ___, ___, ___, ___, ___, ___, ___, ___, ___

6. 39, ___, ___, ___, ___, ___, ___, ___, ___, ___

7. 56, ___, ___, ___, ___, ___, ___, ___, ___, ___

8. 99, ___, ___, ___, ___, ___, ___, ___, ___, ___

▶ **Mixed Review**

Write < , >, or = in the circle.

9. 14 ◯ 19   51 ◯ 51   25 ◯ 30   43 ◯ 40

10. 16 ◯ 16   81 ◯ 71   10 ◯ 15   50 ◯ 42

Problem Solving • Use a Model

Use the model ⬚⬚⬚⬚⬚⬚⬚⬚ ▢.
Find 10 more or 10 less.
Write the number.

1. Ella has 50 blocks.
Cara has 10 less. How
many blocks does
Cara have?

   __40__ blocks

To find
10 less,
cross out
1 ten.

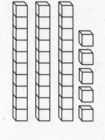

---

2. Arthur has 35 crayons.
Rachel has 10 less.
How many crayons
does Rachel have?

   _____ crayons

To find
10 less,
cross out
1 ten.

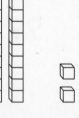

---

3. Steve has 22 toy cars.
Liz has 10 more cars
than Steve. How many
cars does Liz have?

   _____ cars

To find
10 more,
add 1 ten.

---

4. Emma jumps rope 30
times. Charlie jumps rope
10 more times than Emma.
How many times does
Charlie jump rope?

   _____ times

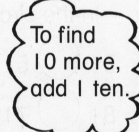

To find
10 more,
add 1 ten.

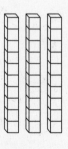

Skip Count by 2s, 5s, and 10s

Skip count.
Write how many.

1.

2   4   6   8   10

2.

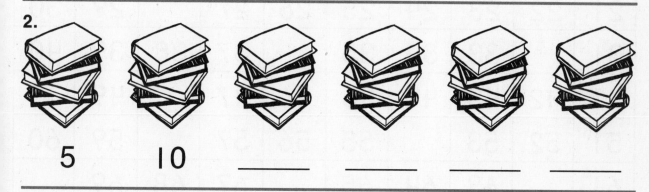

5   10   ___   ___   ___   ___

3.

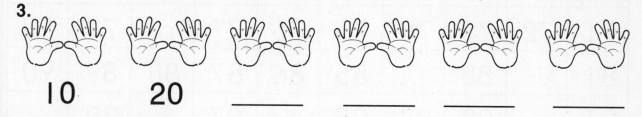

10   20   ___   ___   ___   ___

▶ Mixed Review

Circle the number that is less.

4. 89 or 98   54 or 34   39 or 36

5. 44 or 34   83 or 92   43 or 44

6. 61 or 63   48 or 24   26 or 62

Name _____

Algebra: Use a Hundred Chart to Skip Count

1. Write the missing numbers.
   Count by twos. Color the twos [red ▷ .
   Count by tens. Color the tens [blue ▷ .
   Some boxes will have 2 colors.

1	2	3	4	5	6	7	8	9	
11		13	14	15		17	18	19	
21	22	23	24	25	26	27		29	30
31		33	34	35		37	38	39	40
41	42	43	44	45	46	47		49	
51	52	53		55	56	57		59	60
61		63	64	65		67	68	69	
71	72	73	74	75	76	77		79	80
81		83		85	86	87	88	89	90
91		93	94	95		97		99	

▶ Mixed Review

Circle the greater number.

2. 28 or 18   65 or 56   89 or 98

3. 16 or 60   23 or 13   47 or 42

4. 20 or 22   79 or 76   23 or 32

Algebra: Patterns on a Hundred Chart

Start on the given number.
Count forward by tens.
Write the numbers you say.

Use a hundred chart if you
need to.

1	2	3	4	5	6	7	8	9	10
11	12	13	14	15	16	17	18	19	20
21	22	23	24	25	26	27	28	29	30
31	32	33	34	35	36	37	38	39	40
41	42	43	44	45	46	47	48	49	50
51	52	53	54	55	56	57	58	59	60
61	62	63	64	65	66	67	68	69	70
71	72	73	74	75	76	77	78	79	80
81	82	83	84	85	86	87	88	89	90
91	92	93	94	95	96	97	98	99	100

1. Start on 4.

    14 , 24 , ____ , ____ , ____ , ____ , ____ , ____ , ____

2. Start on 8.

    ____ , ____ , ____ , ____ , ____ , ____ , ____ , ____

3. Start on 6.

    ____ , ____ , ____ , ____ , ____ , ____ , ____ , ____

▶ **Mixed Review**

Count forward. Write the numbers.

4. 31 ____ , ____ , ____ , ____ , ____ , ____ , ____ , ____

5. 63 ____ , ____ , ____ , ____ , ____ , ____ , ____ , ____

Even and Odd

1. Color 51, 53, and 55 | blue |▷.
   Color 52, 54, and 56 | red |▷.
   Color to continue the pattern.

51	52	53	54	55	56	57	58	59	60
61	62	63	64	65	66	67	68	69	70
71	72	73	74	75	76	77	78	79	80
81	82	83	84	85	86	87	88	89	90
91	92	93	94	95	96	97	98	99	100

Write **even** or **odd**.

2. Are the red-shaded numbers even or odd? _____

3. Are the blue-shaded numbers even or odd? _____

4. Start with 60. Skip count by tens.

   Do you say odd or even numbers? _____

5. Start with 52. Skip count by twos.

   Do you say odd or even numbers? _____

▶ **Mixed Review**

Solve.

6. $10 - 3 =$ ____ $9 - 4 =$ ____ $6 - 5 =$ ____

7. $10 - 5 =$ ____ $8 - 3 =$ ____ $7 - 2 =$ ____

8. $5 + 2 =$ ____ $4 + 4 =$ ____ $7 + 2 =$ ____

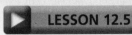

Problem Solving • Find a Pattern

Find a pattern to solve.
Write how many.

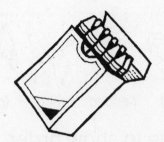

1. There are 10 crayons in each box.
   How many crayons are in 7 boxes?

number of boxes	1	2	3	4	5	6	7
number of crayons	10						

There are _____ crayons in 7 boxes.

2. There are 5 fingers on each hand.
   How many fingers are on 6 hands?

number of hands	1	2	3	4	5	6
number of fingers	5					

There are _____ fingers on 6 hands.

▶ **Mixed Review**

Write the difference.

3.
$$\begin{array}{r} 7 \\ -3 \\ \hline \end{array} \qquad \begin{array}{r} 9 \\ -1 \\ \hline \end{array} \qquad \begin{array}{r} 8 \\ -4 \\ \hline \end{array} \qquad \begin{array}{r} 5 \\ -0 \\ \hline \end{array} \qquad \begin{array}{r} 3 \\ -3 \\ \hline \end{array}$$

Ordinal Numbers

first second third fourth fifth sixth seventh eighth ninth tenth

Circle to show order.

1.

first

first [red] sixth [green] ninth [yellow]

2.

first

second [red] fifth [green] tenth [yellow]

3.

first

third [red] fourth [blue] fifth [yellow]

4.

first

seventh [red] eighth [blue] ninth [green]

▶ **Mixed Review**

Write >, <, or = in the ◯.

5. 17 ◯ 71 6. 38 ◯ 29 7. 46 ◯ 46

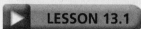

Count On to Add

Circle the greater number.
Use the number line. Count on to add.

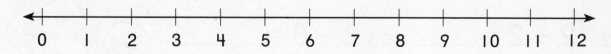

1.
$$\begin{array}{r} 1 \\ + 8 \\ \hline \end{array} \quad \begin{array}{r} 9 \\ + 2 \\ \hline \end{array} \quad \begin{array}{r} 6 \\ + 3 \\ \hline \end{array} \quad \begin{array}{r} 3 \\ + 2 \\ \hline \end{array} \quad \begin{array}{r} 7 \\ + 1 \\ \hline \end{array}$$

2.
$$\begin{array}{r} 9 \\ + 1 \\ \hline \end{array} \quad \begin{array}{r} 2 \\ + 4 \\ \hline \end{array} \quad \begin{array}{r} 10 \\ + 2 \\ \hline \end{array} \quad \begin{array}{r} 5 \\ + 3 \\ \hline \end{array} \quad \begin{array}{r} 9 \\ + 3 \\ \hline \end{array}$$

3.
$$\begin{array}{r} 8 \\ + 3 \\ \hline \end{array} \quad \begin{array}{r} 3 \\ + 9 \\ \hline \end{array} \quad \begin{array}{r} 2 \\ + 5 \\ \hline \end{array} \quad \begin{array}{r} 4 \\ + 1 \\ \hline \end{array} \quad \begin{array}{r} 1 \\ + 9 \\ \hline \end{array}$$

4.
$$\begin{array}{r} 2 \\ + 10 \\ \hline \end{array} \quad \begin{array}{r} 7 \\ + 3 \\ \hline \end{array} \quad \begin{array}{r} 10 \\ + 1 \\ \hline \end{array} \quad \begin{array}{r} 3 \\ + 6 \\ \hline \end{array} \quad \begin{array}{r} 5 \\ + 2 \\ \hline \end{array}$$

▶ **Mixed Review**

Solve.

5. $2 + 2 = $ _____ $3 + 1 = $ _____ $4 + 3 = $ _____

6. $7 - 2 = $ _____ $6 - 4 = $ _____ $8 - 4 = $ _____

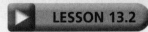
Doubles and Doubles Plus 1

Write the sum.

1. $3 + 3 =$ __6__, so $3 + 4 =$ ____.

2. $2 + 2 =$ ____, so $2 + 3 =$ ____.

3. $4 + 4 =$ ____, so $4 + 5 =$ ____.

4. $1 + 1 =$ ____, so $1 + 2 =$ ____.

5. $0 + 0 =$ ____, so $0 + 1 =$ ____.

6. $5 + 5 =$ ____, so $5 + 6 =$ ____.

$1+1=2$,
so $1+2=3$.

Circle the doubles plus one facts. Then write the sums.

7.
$$\begin{array}{r} 0 \\ +\ 1 \\ \hline \end{array} \qquad \begin{array}{r} 5 \\ +\ 3 \\ \hline \end{array} \qquad \begin{array}{r} 4 \\ +\ 5 \\ \hline \end{array} \qquad \begin{array}{r} 1 \\ +\ 2 \\ \hline \end{array} \qquad \begin{array}{r} 7 \\ +\ 3 \\ \hline \end{array} \qquad \begin{array}{r} 2 \\ +\ 3 \\ \hline \end{array}$$

8.
$$\begin{array}{r} 3 \\ +\ 3 \\ \hline \end{array} \qquad \begin{array}{r} 3 \\ +\ 4 \\ \hline \end{array} \qquad \begin{array}{r} 4 \\ +\ 4 \\ \hline \end{array} \qquad \begin{array}{r} 2 \\ +\ 1 \\ \hline \end{array} \qquad \begin{array}{r} 6 \\ +\ 5 \\ \hline \end{array} \qquad \begin{array}{r} 2 \\ +\ 2 \\ \hline \end{array}$$

▶ **Mixed Review**

Solve.

9. $3 + 4 =$ ____ $2 + 3 =$ ____ $4 + 4 =$ ____

10. $9 - 4 =$ ____ $7 - 2 =$ ____ $6 - 2 =$ ____

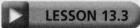

Algebra: Add 3 Numbers

Circle the two numbers you add first. Write the sum.

1.
$$\begin{array}{r} ④ \\ ⑤ \\ +\ 0 \\ \hline 9 \end{array}$$
$$\begin{array}{r} 4 \\ 3 \\ +\ 1 \\ \hline \end{array}$$
$$\begin{array}{r} 6 \\ 6 \\ +\ 0 \\ \hline \end{array}$$
$$\begin{array}{r} 4 \\ 5 \\ +\ 2 \\ \hline \end{array}$$
$$\begin{array}{r} 6 \\ 1 \\ +\ 4 \\ \hline \end{array}$$

2.
$$\begin{array}{r} 5 \\ 3 \\ +\ 2 \\ \hline \end{array}$$
$$\begin{array}{r} 3 \\ 2 \\ +\ 3 \\ \hline \end{array}$$
$$\begin{array}{r} 3 \\ 4 \\ +\ 3 \\ \hline \end{array}$$
$$\begin{array}{r} 2 \\ 5 \\ +\ 5 \\ \hline \end{array}$$
$$\begin{array}{r} 2 \\ 3 \\ +\ 4 \\ \hline \end{array}$$

3.
$$\begin{array}{r} 2 \\ 1 \\ +\ 6 \\ \hline \end{array}$$
$$\begin{array}{r} 3 \\ 3 \\ +\ 4 \\ \hline \end{array}$$
$$\begin{array}{r} 3 \\ 5 \\ +\ 2 \\ \hline \end{array}$$
$$\begin{array}{r} 1 \\ 1 \\ +\ 6 \\ \hline \end{array}$$
$$\begin{array}{r} 2 \\ 5 \\ +\ 1 \\ \hline \end{array}$$

4.
$$\begin{array}{r} 7 \\ 3 \\ +\ 2 \\ \hline \end{array}$$
$$\begin{array}{r} 4 \\ 2 \\ +\ 2 \\ \hline \end{array}$$
$$\begin{array}{r} 1 \\ 3 \\ +\ 2 \\ \hline \end{array}$$
$$\begin{array}{r} 7 \\ 4 \\ +\ 1 \\ \hline \end{array}$$
$$\begin{array}{r} 5 \\ 2 \\ +\ 1 \\ \hline \end{array}$$

▶ **Mixed Review**

Solve.

5. $5 + 5 =$ _____ $8 + 3 =$ _____ $3 + 7 =$ _____

6. $6 + 5 =$ _____ $7 + 4 =$ _____ $2 + 8 =$ _____

7. $2 + 7 =$ _____ $3 + 9 =$ _____ $4 + 8 =$ _____

Problem Solving • Write a Number Sentence

Write a number sentence.
Draw a picture to check.

1. 9 children are at a party.
   3 more children come.
   How many children are at
   the party now?

   __9__ ⊕ __3__ ⊜ __12__ children

2. Lilly has filled two pages of
   her book with stamps.
   Each page has 5 stamps.
   How many stamps does
   she have in all?

   _____ ◯ _____ ◯ _____ stamps

3. Amy has 9 ribbons.
   May gives her 2 more ribbons.
   How many ribbons does she
   have in all?

   _____ ◯ _____ ◯ _____ ribbons

4. Sam walks his dog
   4 times every week.
   How many times does
   he walk it in two weeks?

   _____ ◯ _____ ◯ _____ times

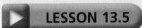

Count Back to Subtract

Use the number line to count back.
Write the difference.

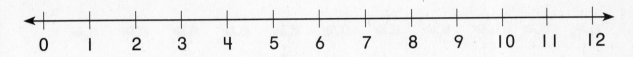

0 1 2 3 4 5 6 7 8 9 10 11 12

1.
```
  12      9      11      7      11      6
 - 3    - 3    - 2    - 2    - 3    - 3
  9
```

2.
```
   7      8      10      9      6      8
 - 1    - 3    - 3    - 2    - 2    - 1
```

3.
```
   6      10      8      7      10      9
 - 1    - 1    - 2    - 3    - 2    - 1
```

▶ **Mixed Review**

Circle the number that is less.

4. 4 or 7	5. 8 or 10	6. 3 or 6
7. 5 or 3	8. 9 or 6	9. 8 or 9
10. 9 or 10	11. 1 or 4	12. 5 or 3

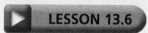

Subtract to Compare

Draw lines to match.
Subtract to find how many more.

1.

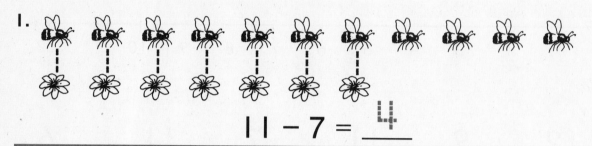

$11 - 7 = \underline{4}$

2.

$10 - 8 = \underline{}$

3.

$11 - 9 = \underline{}$

4.

$12 - 8 = \underline{}$

 Mixed Review

Solve.

5. $5 + 4 = \underline{}$ $3 + 2 = \underline{}$ $7 - 3 = \underline{}$

6. $8 - 3 = \underline{}$ $5 + 2 = \underline{}$ $4 + 2 = \underline{}$

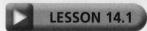

Algebra • Related Addition and Subtraction Facts

Write the sum or difference.
Circle the related facts in each row.

      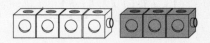

1. $(4 + 3 = \underline{7})$ $3 + 3 = \underline{6}$ $(7 - 3 = \underline{4})$

2. $8 + 3 = \underline{}$ $11 - 3 = \underline{}$ $7 + 4 = \underline{}$

3. $6 + 4 = \underline{}$ $6 - 4 = \underline{}$ $10 - 4 = \underline{}$

4. $7 + 5 = \underline{}$ $3 + 9 = \underline{}$ $12 - 9 = \underline{}$

5. $4 + 8 = \underline{}$ $5 + 6 = \underline{}$ $11 - 6 = \underline{}$

6. $2 + 9 = \underline{}$ $11 - 9 = \underline{}$ $11 - 4 = \underline{}$

▶ **Mixed Review**

Solve.

7. $10 - 8 = \underline{}$ $9 + 2 = \underline{}$ $9 - 5 = \underline{}$

8. $11 - 8 = \underline{}$ $7 + 3 = \underline{}$ $7 - 2 = \underline{}$

9. $10 - 4 = \underline{}$ $8 + 2 = \underline{}$ $12 - 6 = \underline{}$

Fact Families to 12

Add or subtract.
Write the numbers in the fact family.

1.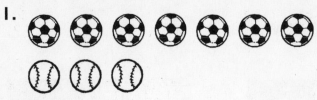

$$\begin{array}{r} 7 \\ + 3 \\ \hline 10 \end{array} \quad \begin{array}{r} 3 \\ + 7 \\ \hline 10 \end{array} \quad \begin{array}{r} 10 \\ - 3 \\ \hline 7 \end{array} \quad \begin{array}{r} 10 \\ - 7 \\ \hline 3 \end{array}$$

[7] [3] [10]

2.

$$\begin{array}{r} 4 \\ + 7 \\ \hline \end{array} \quad \begin{array}{r} 7 \\ + 4 \\ \hline \end{array} \quad \begin{array}{r} 11 \\ - 7 \\ \hline \end{array} \quad \begin{array}{r} 11 \\ - 4 \\ \hline \end{array}$$

[ ] [ ] [ ]

3.

$$\begin{array}{r} 8 \\ + 4 \\ \hline \end{array} \quad \begin{array}{r} 4 \\ + 8 \\ \hline \end{array} \quad \begin{array}{r} 12 \\ - 4 \\ \hline \end{array} \quad \begin{array}{r} 12 \\ - 8 \\ \hline \end{array}$$

[ ] [ ] [ ]

▶ **Mixed Review**

Solve.

4. $10 - 9 =$ ___

5. $8 - 4 =$ ___

6. $3 + 4 + 5 =$ ___

7. $6 + 3 + 2 =$ ___

Sums and Differences to 12

Write the sum or difference.

1. $\begin{array}{r} 7 \\ + 3 \\ \hline 10 \end{array}$	$\begin{array}{r} 5 \\ + 5 \\ \hline \end{array}$	$\begin{array}{r} 10 \\ - 3 \\ \hline \end{array}$	$\begin{array}{r} 10 \\ - 5 \\ \hline \end{array}$	$\begin{array}{r} 7 \\ + 5 \\ \hline \end{array}$	$\begin{array}{r} 12 \\ - 5 \\ \hline \end{array}$
2. $\begin{array}{r} 8 \\ - 3 \\ \hline \end{array}$	$\begin{array}{r} 5 \\ + 3 \\ \hline \end{array}$	$\begin{array}{r} 6 \\ + 3 \\ \hline \end{array}$	$\begin{array}{r} 9 \\ - 3 \\ \hline \end{array}$	$\begin{array}{r} 6 \\ + 4 \\ \hline \end{array}$	$\begin{array}{r} 4 \\ + 5 \\ \hline \end{array}$
3. $\begin{array}{r} 5 \\ + 6 \\ \hline \end{array}$	$\begin{array}{r} 9 \\ - 2 \\ \hline \end{array}$	$\begin{array}{r} 10 \\ + 2 \\ \hline \end{array}$	$\begin{array}{r} 12 \\ - 6 \\ \hline \end{array}$	$\begin{array}{r} 8 \\ - 4 \\ \hline \end{array}$	$\begin{array}{r} 11 \\ - 3 \\ \hline \end{array}$
4. $\begin{array}{r} 9 \\ + 1 \\ \hline \end{array}$	$\begin{array}{r} 3 \\ + 5 \\ \hline \end{array}$	$\begin{array}{r} 8 \\ - 6 \\ \hline \end{array}$	$\begin{array}{r} 8 \\ - 5 \\ \hline \end{array}$	$\begin{array}{r} 10 \\ - 1 \\ \hline \end{array}$	$\begin{array}{r} 11 \\ + 1 \\ \hline \end{array}$

▶ **Mixed Review**

Draw a line under the pattern unit.
Then draw shapes to continue.
Complete the pattern.

5. ▭ △ ◯ ▭ △ ◯ ▭ ___ ___

6. ▢ ▢ △ ▢ ▢ △ ▢ ___ ___

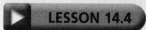

Algebra • Missing Numbers

Write the missing number.
Use 🎲 if you need to.

1. $3 + \boxed{2} = 5$ $5 - 3 = \boxed{2}$

2. $\boxed{} + 4 = 8$ $8 - 4 = \boxed{}$

3. $6 + \boxed{} = 11$ $11 - 6 = \boxed{}$

4. $9 + \boxed{} = 12$ $12 - 9 = \boxed{}$

5. $\boxed{} + 5 = 6$ $6 - 5 = \boxed{}$

6. $4 + \boxed{} = 10$ $10 - 4 = \boxed{}$

▶ **Mixed Review**

Solve.
Then write two related subtraction facts.

7. $10 + 2 = \boxed{}$

$\boxed{} - \boxed{} = 10$

$\boxed{} - \boxed{} = 2$

8. $5 + 6 = \boxed{}$

$\boxed{} - \boxed{} = 5$

$\boxed{} - \boxed{} = 6$

9. $4 + 5 = \boxed{}$

$\boxed{} - \boxed{} = 4$

$\boxed{} - \boxed{} = 5$

Problem Solving • Choose a Strategy

Choose a way to solve each
problem. Make a model 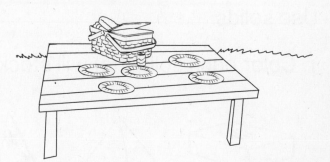,
draw a picture ▭▷,
or write a number sentence
✎ . Show your work.

1. 7 children sit at the table.
Then 3 more come.
How many children are at
the table?

   ___10___ children

2. There are 7 blue cups
and 5 red cups.
How many cups are
there in all?

   _____ cups

3. There are 12 bugs.
Then 3 of them
crawl away. How
many bugs are left?

   _____ bugs

4. 5 ducks swim in the pond.
5 more ducks jump in.
How many ducks are
in the pond?

   _____ ducks

Solid Figures

Use solids.

1. Color each solid that will stack.

2. Color each solid that will roll.

3. Color each solid that will slide.

▶ **Mixed Review**

Write the missing number.

4. 77, ___, 79   44, 45, ___   ___, 32, 33

5. 56, 57, ___   88, ___, 90   ___, 91, 92

Name _____

Faces and Vertices

Use solids.
Color the pictures that match the sentence.

1. This solid has 5 faces.

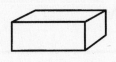 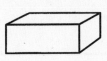

2. These solids have 6 faces.

3. These solids have 8 vertices.

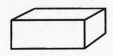 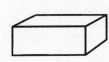

4. This solid has 5 vertices.

▶ **Mixed Review**

Solve.

5. $6 + 3 =$ ____ $4 + 4 =$ ____ $7 - 3 =$ ____

6. $8 - 5 =$ ____ $3 + 2 =$ ____ $6 + 2 =$ ____

Plane Shapes on Solid Figures

Draw a building. Use at least one ▭, one ◯, one △, and one ☐.

1. Color ▭ blue.   **2.** Color ◯ green.

3. Color △ red.   **4.** Color ☐ yellow.

▶ **Mixed Review**

Write + or − in the circle.

5. 2 ◯ 5 = 7   5 ◯ 3 = 2   5 ◯ 4 = 9

6. 8 ◯ 4 = 4   7 ◯ 2 = 5   6 ◯ 5 = 1

Sort and Identify Plane Shapes

Use ▐ **blue** ▐ ▷ to trace each side.

Use ▐ **red** ▐ ▷ to circle each vertex.

Write how many sides and vertices there are.

1.

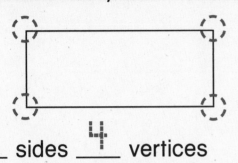

__4__ sides __4__ vertices

2.

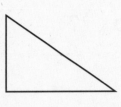

____ sides ____ vertices

3.

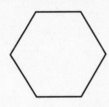

____ sides ____ vertices

4.

____ sides ____ vertices

5.

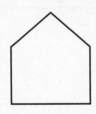

____ sides ____ vertices

6.

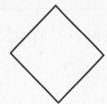

____ sides ____ vertices

▶ Mixed Review

Circle the number that is greater.

7. 5 or 7   4 or 2   8 or 9

8. 6 or 5   1 or 7   6 or 4

Problem Solving • Make a Model

Use pattern blocks to make a model.
Draw to show your model.
Write how many pattern blocks you use.

1. How many ▱
   make a ⬡?

   ____ ▱

2. How many △
   make a ⬡?

   ____ △

3. How many ▱ and △
   make a ▱?

   ____ ▱ ____ △

4. How many △ and ▱
   make a ⬡?

   ____ △ ____ ▱

Open and Closed

Circle each open figure.
Color each closed figure.

1.

2.

3.

4.

5.

6.

7.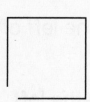

8.

9.

▶ **Mixed Review**

Trace each side.
Circle each vertex.
Write how many sides and vertices there are.

10.

____ sides ____ vertices

11.

____ sides ____ vertices

Problem Solving • Use a Picture

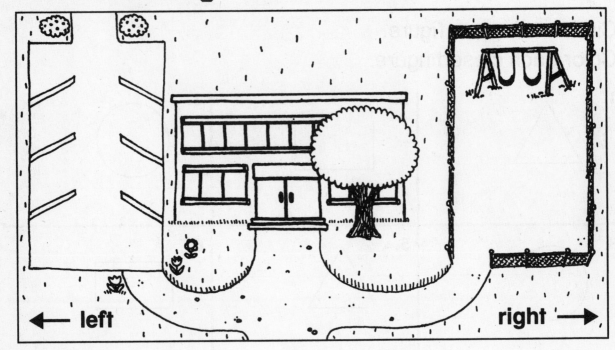

Follow the directions.

1. The 🌳 is to the right of the 🚪.

   Draw a 📬 to the right of the 🌳.

2. The 🌼 are to the left of the 🚪.

   Draw a 🚗 to the left of the 🏢.

3. The ⛓ is in front of the ▨.

   Draw a 🚲 in front of the ⛓.

4. The ▯ is above the 🚪.

   Draw a 🐦 above the 🏢.

5. The 🌼 are next to the 🌾.

   Draw more 🌼 next to the 🌾.

6. The ⛓ is near the ▨.

   Draw a 🛝 near the ▨.

Give and Follow Directions

Follow the directions in order.
Draw the path. Write the place.

1. Go **right** 4.
   Go **up** 2.
   Where are you?

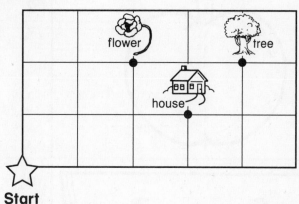

Start

2. Go **down** 1.
   Go **left** 3.
   Go **down** 1.
   Where are you?

Start

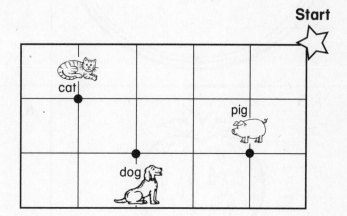

▶ **Mixed Review**

Write the difference.

3. $9 - 4 =$ _____ $5 - 4 =$ _____ $10 - 6 =$ _____

4. $7 - 3 =$ _____ $8 - 5 =$ _____ $9 - 6 =$ _____

Symmetry

Draw a line of symmetry to show two matching parts.

1.
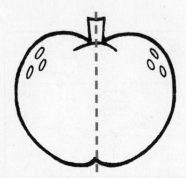

2.
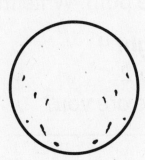

3.

4.

5.

6.

▶ **Mixed Review**

Write the sum.

7. $2 + 4 =$ _____ $4 + 3 =$ _____ $5 + 4 =$ _____

8. $4 + 1 =$ _____ $6 + 2 =$ _____ $3 + 6 =$ _____

Slides and Turns

Circle **slide** or **turn** to name the move.

1.

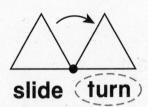

slide (turn)

2.

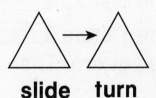

slide   turn

3.

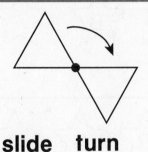

slide   turn

4.

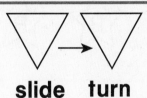

slide   turn

5.

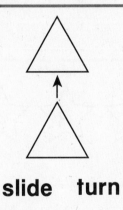

slide   turn

6.

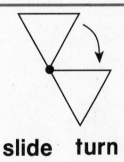

slide   turn

▶ **Mixed Review**

7. Circle each solid that will roll.

8. Circle each solid that will stack.

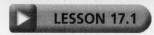

Algebra: Describe and Extend Patterns

Color the R stars ▮ red ▷.
Color the B stars ▮ blue ▷.
Color the Y stars ▮ yellow ▷.
Find the pattern. Then color to continue it.

1.

R B R B R B ☆ ☆

2.

B R R B R R B ☆ ☆

3.

R B Y R B Y R ☆ ☆

4.

B B Y B B Y B ☆ ☆

▶ **Mixed Review**

Complete the doubles facts.

5. $3 + \underline{3} = \underline{6}$ $6 + \underline{} = \underline{}$ $2 + \underline{} = \underline{}$

6. $4 + \underline{} = \underline{}$ $5 + \underline{} = \underline{}$ $1 + \underline{} = \underline{}$

Algebra: Pattern Units

Circle the pattern unit.

1.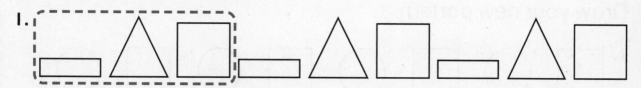

2.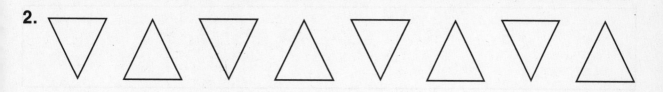

3.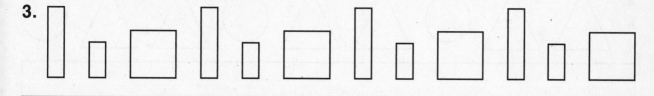

4.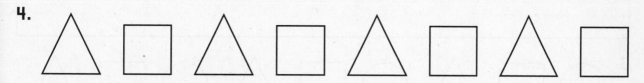

5.

► **Mixed Review**

Circle the number that is less.

6. 13 or 11   10 or 8   9 or 10

7. 14 or 18   3 or 5   6 or 4

Algebra: Make New Patterns

Use the same shapes to make a different pattern.
Draw your new pattern.

1.

2.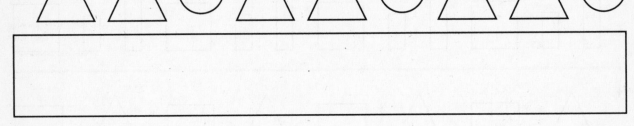

3.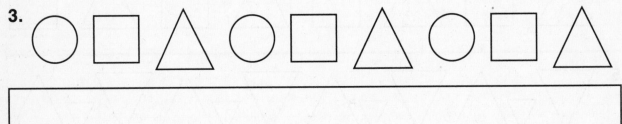

▶ **Mixed Review**

Add.

4. $3 + 5 =$ ___ $8 + 2 =$ ___ $3 + 4 =$ ___

5. $4 + 2 =$ ___ $5 + 2 =$ ___ $2 + 2 =$ ___

Problem Solving • Correct a Pattern

Find the pattern.
Circle the mistake.
Draw the correct shape.

Each pattern unit
is 3 shapes long.

1.
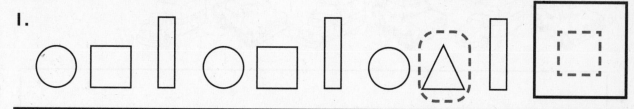

2.
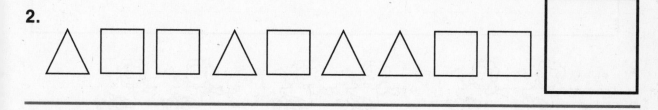

3.

4.

5.
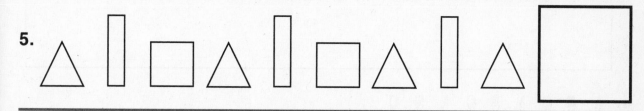

6.

Problem Solving • Transfer Patterns

Use shapes to show the same pattern.
Draw the shapes.

1.

2.

3.

4.

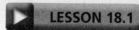

Doubles and Doubles Plus 1

Write the sums.

1. $6 + 6 =$ __12__,   so $6 + 7 =$ __13__

2. $9 + 9 =$ ____,   so $9 + 10 =$ ____

3. $7 + 7 =$ ____,   so $8 + 7 =$ ____

4. $5 + 5 =$ ____,   so $5 + 6 =$ ____

5. $8 + 8 =$ ____,   so $9 + 8 =$ ____

6. $4 + 4 =$ ____,   so $5 + 4 =$ ____

Write the sums.

7.
$$\begin{array}{cccccc} 10 & 6 & 5 & 8 & 7 & 9 \\ +10 & +6 & +5 & +8 & +7 & +9 \\ \hline \end{array}$$

8.
$$\begin{array}{cccccc} 5 & 8 & 5 & 10 & 7 & 6 \\ +6 & +9 & +4 & +9 & +8 & +7 \\ \hline \end{array}$$

▶ **Mixed Review**

Add.

9. $6 + 6 =$ ___ $4 + 4 =$ ___ $2 + 2 =$ ___

10. $3 + 3 =$ ___ $1 + 1 =$ ___ $5 + 5 =$ ___

Name _____

10 and More

Write the sum.

1.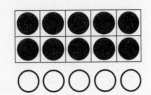
$$\begin{array}{r} 10 \\ +\ 3 \\ \hline 13 \end{array}$$

2.
$$\begin{array}{r} 10 \\ +\ 8 \\ \hline \end{array}$$

3.
$$\begin{array}{r} 10 \\ +\ 5 \\ \hline \end{array}$$

4.
$$\begin{array}{r} 10 \\ +\ 7 \\ \hline \end{array}$$

5.
$$\begin{array}{r} 10 \\ +\ 2 \\ \hline \end{array} \qquad \begin{array}{r} 6 \\ +10 \\ \hline \end{array} \qquad \begin{array}{r} 10 \\ +\ 0 \\ \hline \end{array} \qquad \begin{array}{r} 10 \\ +\ 4 \\ \hline \end{array} \qquad \begin{array}{r} 9 \\ +10 \\ \hline \end{array} \qquad \begin{array}{r} 10 \\ +\ 1 \\ \hline \end{array}$$

6.
$$\begin{array}{r} 5 \\ +10 \\ \hline \end{array} \qquad \begin{array}{r} 10 \\ +\ 8 \\ \hline \end{array} \qquad \begin{array}{r} 10 \\ +\ 9 \\ \hline \end{array} \qquad \begin{array}{r} 10 \\ +\ 3 \\ \hline \end{array} \qquad \begin{array}{r} 2 \\ +10 \\ \hline \end{array} \qquad \begin{array}{r} 10 \\ +\ 7 \\ \hline \end{array}$$

▶ **Mixed Review**

Solve.

7. $10 - 6 =$ _____ $10 + 5 =$ _____ $10 - 5 =$ _____

8. $10 + 6 =$ _____ $10 - 7 =$ _____ $10 + 7 =$ _____

9. $8 + 8 =$ _____ $9 + 8 =$ _____ $9 - 8 =$ _____

Make 10 to Add

Use ● and Workmat 7. Show the numbers and add.
Then make a ten and add.

1.

$$\begin{array}{r} 9 \\ + 7 \\ \hline 16 \end{array}$$

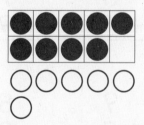

$$\begin{array}{r} 10 \\ + 6 \\ \hline 16 \end{array}$$

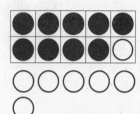

2.

$$\begin{array}{r} 9 \\ + 3 \\ \hline \end{array}$$

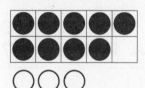

$$\begin{array}{r} 10 \\ + 2 \\ \hline \end{array}$$

3.

$$\begin{array}{r} 9 \\ + 6 \\ \hline \end{array}$$

$$\begin{array}{r} 10 \\ + 5 \\ \hline \end{array}$$

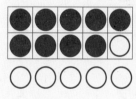

Add.

4.

$$\begin{array}{r} 9 \\ + 5 \\ \hline \end{array} \qquad \begin{array}{r} 9 \\ + 8 \\ \hline \end{array} \qquad \begin{array}{r} 9 \\ + 2 \\ \hline \end{array} \qquad \begin{array}{r} 9 \\ + 9 \\ \hline \end{array} \qquad \begin{array}{r} 9 \\ + 4 \\ \hline \end{array} \qquad \begin{array}{r} 9 \\ + 7 \\ \hline \end{array}$$

▶ **Mixed Review**

Write the number that is between.

5. 64, ____, 66   44, ____, 46   25, ____, 27

6. 47, ____, 49   5, ____, 7   53, ____, 55

7. 22, ____, 24   36, ____, 38   31, ____, 33

Use Make a 10

Use ● and Workmat 7 to add.
Start with the greater number.

1.   7
      + 6
      1 3

Think:
7 + 6 = 10 + 3

2.   4   8   6   9   5   9
    + 9   + 3   + 7   + 8   + 7   + 3

3.   8   9   6   5   7   8
    + 5   + 2   + 8   + 6   + 9   + 4

4.   7   6   8   9   8   4
    + 8   + 9   + 6   + 5   + 9   + 7

▶ **Mixed Review**

Find the sum.

5. 4 + 4 = ____   9 + 9 = ____   2 + 2 = ____

6. 3 + 3 = ____   1 + 1 = ____   5 + 5 = ____

7. 6 + 6 = ____   7 + 7 = ____   8 + 8 = ____

Algebra: Add 3 Numbers

Circle the numbers you add first.
Write the sum.

1.
$$\begin{array}{r} ⑦ \\ ⑦ \\ + 2 \\ \hline 16 \end{array}$$
$$\begin{array}{r} 5 \\ 5 \\ + 1 \\ \hline \end{array}$$
$$\begin{array}{r} 2 \\ 8 \\ + 6 \\ \hline \end{array}$$
$$\begin{array}{r} 9 \\ 2 \\ + 1 \\ \hline \end{array}$$
$$\begin{array}{r} 1 \\ 8 \\ + 8 \\ \hline \end{array}$$

2.
$$\begin{array}{r} 3 \\ 3 \\ + 6 \\ \hline \end{array}$$
$$\begin{array}{r} 3 \\ 7 \\ + 2 \\ \hline \end{array}$$
$$\begin{array}{r} 1 \\ 6 \\ + 9 \\ \hline \end{array}$$
$$\begin{array}{r} 6 \\ 6 \\ + 3 \\ \hline \end{array}$$
$$\begin{array}{r} 7 \\ 5 \\ + 5 \\ \hline \end{array}$$

3.
$$\begin{array}{r} 1 \\ 5 \\ + 9 \\ \hline \end{array}$$
$$\begin{array}{r} 7 \\ 8 \\ + 3 \\ \hline \end{array}$$
$$\begin{array}{r} 5 \\ 6 \\ + 4 \\ \hline \end{array}$$
$$\begin{array}{r} 8 \\ 5 \\ + 2 \\ \hline \end{array}$$
$$\begin{array}{r} 4 \\ 4 \\ + 8 \\ \hline \end{array}$$

▶ **Mixed Review**

Skip-count. Write the missing number.

4. 26, 28, 30, _____, 34

5. 20, _____, 40, 50, 60

6. 55, 60, 65, _____, 75

7. 12, 14, _____, 18, 20

8. 40, 50, _____, 70, 80

9. 30, 35, 40, _____, 50

10. 25, 30, _____, 40, 45

11. 44, 46, 48, _____, 52

Problem Solving • Use Data from a Table

This table tells how many leaves children found.

Use the table to answer the questions.

Leaves		Number
oak		1
maple		3
dogwood		2
birch		6

1. How many dogwood and oak leaves did they find in all?

   __3__ leaves

   Think:
   Find the numbers and add.
   __2__ $\oplus$ __1__ $\ominus$ __3__

2. How many more birch leaves than maple leaves did they find?

   _____ more birch leaves

   Think:
   Find the numbers and subtract.
   ___ $\bigcirc$ ___ $\bigcirc$ ___

3. How many birch and maple leaves did they find in all?

   _____ leaves

   ___ $\bigcirc$ ___ $\bigcirc$ ___

4. How many more maple leaves than oak leaves did they find?

   _____ more maple leaves

   ___ $\bigcirc$ ___ $\bigcirc$ ___

5. How many leaves were not oak or maple?

   _____ leaves

   ___ $\bigcirc$ ___ $\bigcirc$ ___

Use a Number Line to Count Back

$$\begin{array}{r} 11 \\ -\ 3 \\ \hline 8 \end{array}$$

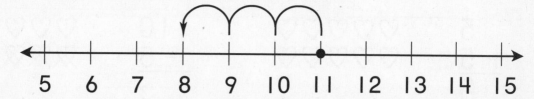

5   6   7   8   9   10   11   12   13   14   15

Count back to subtract. Write the difference.
Use the number line to help.

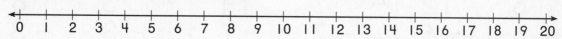

0 1 2 3 4 5 6 7 8 9 10 11 12 13 14 15 16 17 18 19 20

1.

$$\begin{array}{r} 10 \\ -\ 1 \\ \hline 9 \end{array} \qquad \begin{array}{r} 9 \\ -\ 2 \\ \hline \end{array} \qquad \begin{array}{r} 11 \\ -\ 2 \\ \hline \end{array} \qquad \begin{array}{r} 7 \\ -\ 3 \\ \hline \end{array} \qquad \begin{array}{r} 10 \\ -\ 3 \\ \hline \end{array} \qquad \begin{array}{r} 12 \\ -\ 3 \\ \hline \end{array}$$

Use the number line to subtract.

2.

$$\begin{array}{r} 18 \\ -\ 9 \\ \hline \end{array} \qquad \begin{array}{r} 17 \\ -\ 8 \\ \hline \end{array} \qquad \begin{array}{r} 15 \\ -\ 6 \\ \hline \end{array} \qquad \begin{array}{r} 13 \\ -\ 7 \\ \hline \end{array} \qquad \begin{array}{r} 10 \\ -\ 9 \\ \hline \end{array} \qquad \begin{array}{r} 14 \\ -\ 8 \\ \hline \end{array}$$

3.

$$\begin{array}{r} 10 \\ -\ 2 \\ \hline \end{array} \qquad \begin{array}{r} 16 \\ -\ 9 \\ \hline \end{array} \qquad \begin{array}{r} 12 \\ -\ 4 \\ \hline \end{array} \qquad \begin{array}{r} 14 \\ -\ 5 \\ \hline \end{array} \qquad \begin{array}{r} 16 \\ -\ 7 \\ \hline \end{array} \qquad \begin{array}{r} 12 \\ -\ 5 \\ \hline \end{array}$$

▶ **Mixed Review**

Skip-count. Write the missing number.

4. 20, 25, _____, 35

5. 4, 6, _____, 10

6. 28, 30, _____, 34

7. 5, 10, _____, 20

8. 30, 40, _____, 60

9. 20, 30, _____, 50

Name _____

Doubles Fact Families

$$5 + 5 = 10 \quad \heartsuit\heartsuit\heartsuit\heartsuit\heartsuit \; \heartsuit\heartsuit\heartsuit\heartsuit \qquad 10 - 5 = 5$$

Write the sum and difference for each pair.

1. $6 + 6$ $12 - 6$
2. $7 + 7$ $14 - 7$
3. $3 + 3$ $6 - 3$

4. $2 + 2$ $4 - 2$
5. $9 + 9$ $18 - 9$
6. $8 + 8$ $16 - 8$

7. $8 - 4$ $4 + 4$
8. $4 - 2$ $2 + 2$
9. $10 - 5$ $5 + 5$

▶ **Mixed Review**

Write the sum.

10. $9 + 6 = \underline{\quad}$ $5 + 7 = \underline{\quad}$ $6 + 7 = \underline{\quad}$

11. $6 + 9 = \underline{\quad}$ $8 + 6 = \underline{\quad}$ $7 + 6 = \underline{\quad}$

12. $7 + 5 = \underline{\quad}$ $6 + 8 = \underline{\quad}$ $9 + 9 = \underline{\quad}$

Algebra • Related Addition and Subtraction Facts

Write the sum and difference for each pair.

1.			2.			3.		
9 + 9 18	18 − 9 9		8 + 6	14 − 6		6 + 6	12 − 6	
4.			5.			6.		
8 + 3	11 − 3		9 + 4	13 − 4		7 + 6	13 − 6	
7.			8.			9.		
9 +7	16 − 7		6 + 5	11 − 5		6 +8	14 − 8	
10.			11.			12.		
6 +9	15 − 9		9 +3	12 − 3		6 +7	13 −7	

▶ **Mixed Review**

Write the sum or difference.

13. $6 + 5 =$ ___ $7 + 3 =$ ___ $4 + 8 =$ ___

14. $11 - 6 =$ ___ $10 - 7 =$ ___ $12 - 4 =$ ___

15. $11 - 5 =$ ___ $10 - 3 =$ ___ $12 - 8 =$ ___

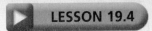

Problem Solving • Estimate Reasonable Answers

Circle the best estimate.

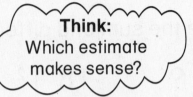

Think: Which estimate makes sense?

1. Dan digs up 12 potatoes.
Rose digs up 9 potatoes.
About how many potatoes
do they dig up in all?

    about 5   about 10   (about 20)

2. Molly picks 4 tomatoes.
Anna picks 5 tomatoes.
About how many tomatoes
do they pick in all?

    about 5   about 10   about 20

3. Amelia picks 10 peaches.
She eats 4 of them.
About how many peaches
does she have now?

    about 5   about 10   about 15

4. Shiwani plants 20 pumpkin plants.
Bugs eat 6 of them.
About how many pumpkin plants
does Shiwani have left?

    about 5   about 10   about 15

Practice the Facts

Add or subtract. Color each sum **green**.

Color each difference **purple**.

1.

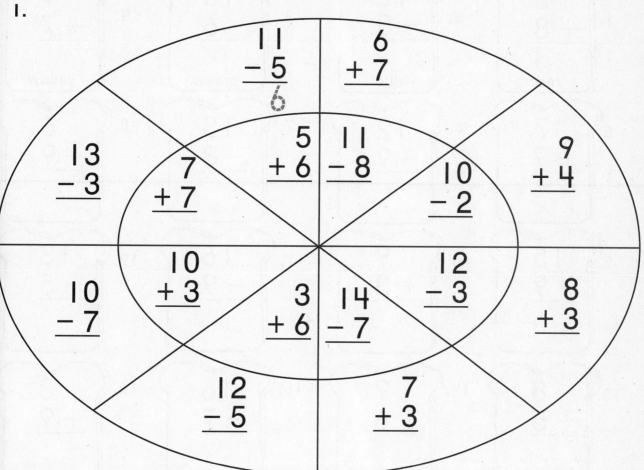

$$\begin{array}{r} 11 \\ -5 \\ \hline 6 \end{array}$$

$$\begin{array}{r} 6 \\ +7 \\ \hline \end{array}$$

$$\begin{array}{r} 13 \\ -3 \\ \hline \end{array}$$

$$\begin{array}{r} 7 \\ +7 \\ \hline \end{array}$$

$$\begin{array}{r} 5 \\ +6 \\ \hline \end{array}$$

$$\begin{array}{r} 11 \\ -8 \\ \hline \end{array}$$

$$\begin{array}{r} 10 \\ -2 \\ \hline \end{array}$$

$$\begin{array}{r} 9 \\ +4 \\ \hline \end{array}$$

$$\begin{array}{r} 10 \\ -7 \\ \hline \end{array}$$

$$\begin{array}{r} 10 \\ +3 \\ \hline \end{array}$$

$$\begin{array}{r} 3 \\ +6 \\ \hline \end{array}$$

$$\begin{array}{r} 14 \\ -7 \\ \hline \end{array}$$

$$\begin{array}{r} 12 \\ -3 \\ \hline \end{array}$$

$$\begin{array}{r} 8 \\ +3 \\ \hline \end{array}$$

$$\begin{array}{r} 12 \\ -5 \\ \hline \end{array}$$

$$\begin{array}{r} 7 \\ +3 \\ \hline \end{array}$$

▶ **Mixed Review**

Solve.

2. $1 + 2 + 7 =$ ___ $2 + 4 + 4 =$ ___

3. $5 + 2 + 5 =$ ___ $6 + 3 + 5 =$ ___

4. $3 + 6 + 3 =$ ___ $6 + 2 + 4 =$ ___

Fact Families to 20

Write the sum or difference.
Color all the facts in the same fact family to match.

1.
$$\begin{array}{r} 5 \\ +\ 8 \\ \hline 13 \end{array}$$
red

2.
$$\begin{array}{r} 12 \\ -\ 3 \\ \hline \end{array}$$
blue

3.
$$\begin{array}{r} 15 \\ -\ 6 \\ \hline \end{array}$$
green

4.
$$\begin{array}{r} 9 \\ +\ 7 \\ \hline \end{array}$$
yellow

5.
$$\begin{array}{r} 7 \\ +\ 9 \\ \hline \end{array}$$

6.
$$\begin{array}{r} 12 \\ -\ 9 \\ \hline \end{array}$$

7.
$$\begin{array}{r} 13 \\ -\ 8 \\ \hline \end{array}$$

8.
$$\begin{array}{r} 6 \\ +\ 9 \\ \hline \end{array}$$

9.
$$\begin{array}{r} 15 \\ -\ 9 \\ \hline \end{array}$$

10.
$$\begin{array}{r} 9 \\ +\ 3 \\ \hline \end{array}$$

11.
$$\begin{array}{r} 16 \\ -\ 9 \\ \hline \end{array}$$

12.
$$\begin{array}{r} 13 \\ -\ 5 \\ \hline \end{array}$$

13.
$$\begin{array}{r} 8 \\ +\ 5 \\ \hline \end{array}$$

14.
$$\begin{array}{r} 9 \\ +\ 6 \\ \hline \end{array}$$

15.
$$\begin{array}{r} 16 \\ -\ 7 \\ \hline \end{array}$$

16.
$$\begin{array}{r} 3 \\ +\ 9 \\ \hline \end{array}$$

▶ **Mixed Review**

Solve.

17. $11 - 7 =$ ___ $15 + 3 =$ ___ $9 + 9 =$ ___

18. $20 - 3 =$ ___ $8 + 8 =$ ___ $16 + 2 =$ ___

19. $7 + 7 =$ ___ $12 - 8 =$ ___ $19 - 4 =$ ___

Algebra: Ways to Make Numbers to 20

Circle all the ways to make the number at the top.

1.

12
8 + 2 + 2
5 + 4 + 3
6 + 7
14 − 2
10 + 2

2.

18
9 + 9
7 + 3 + 2
9 + 5 + 4
20 − 3
3 + 6 + 9

3.

13
9 + 2
7 + 5 + 1
7 + 6
19 − 6
4 + 5 + 5

4.

19
10 + 1 + 8
20 − 1
9 + 6 + 4
9 + 9
8 + 9 + 1

▶ **Mixed Review**

Write + or −. Make each equal 15.

5. 14 1   7 ◯ 8   19 ◯ 4   5 ◯ 5 ◯ 5

Name _____

Problem Solving • Make a Model

Use Workmat 1, 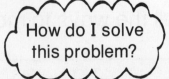, and ⬜.
Draw the 🎲 and ⬜ you use.
Write the answer.

How do I solve this problem?

1. 8 dogs pull a sled. 4 dogs rest. How many dogs are there in all? _____ dogs	
2. 9 penguins are swimming. There are 17 penguins in all. How many penguins are not swimming? _____ penguins	
3. Ella makes 6 snowballs. Louis makes 5 snowballs. How many snowballs do they make in all? _____ snowballs	
4. There are 13 red hats. There are 20 hats in all. How many of the hats are not red? _____ hats	

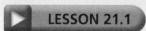

Halves

Find the shapes that show halves. Color $\frac{1}{2}$.

1.

 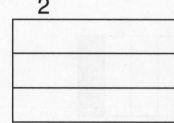

2.

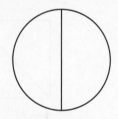 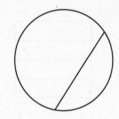 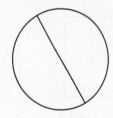

3.

 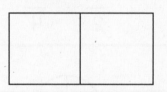 

4.

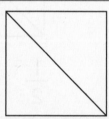 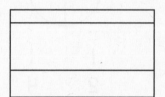 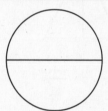

Mixed Review

Circle the number that is less.

5. 27 35	48 58	19 16
6. 37 82	71 67	64 46
7. 92 81	77 61	91 89
8. 23 63	40 82	54 45

Fourths

Color one part. Circle the fraction.

1.

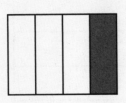

$\dfrac{1}{2}$ $\boxed{\dfrac{1}{4}}$

$\dfrac{1}{2}$ $\dfrac{1}{4}$

$\dfrac{1}{2}$ $\dfrac{1}{4}$

2.

$\dfrac{1}{2}$ $\dfrac{1}{4}$

$\dfrac{1}{2}$ $\dfrac{1}{4}$

$\dfrac{1}{2}$ $\dfrac{1}{4}$

3.

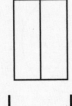

$\dfrac{1}{2}$ $\dfrac{1}{4}$

$\dfrac{1}{2}$ $\dfrac{1}{4}$

$\dfrac{1}{2}$ $\dfrac{1}{4}$

▶ **Mixed Review**

Add or subtract.

4. $3 + 5 =$ _____ $6 + 6 =$ _____ $6 + 5 =$ _____

5. $9 - 2 =$ _____ $12 - 3 =$ _____ $4 + 4 =$ _____

6. $5 + 7 =$ _____ $7 - 3 =$ _____ $11 - 4 =$ _____

Thirds

Color one part. Circle the fraction.

1.

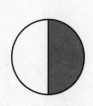   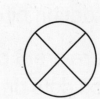

$\frac{1}{3}$ $\frac{1}{2}$ $\frac{1}{4}$ | $\frac{1}{3}$ $\frac{1}{2}$ $\frac{1}{4}$ | $\frac{1}{3}$ $\frac{1}{2}$ $\frac{1}{4}$

2.

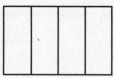   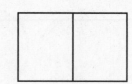

$\frac{1}{3}$ $\frac{1}{2}$ $\frac{1}{4}$ | $\frac{1}{3}$ $\frac{1}{2}$ $\frac{1}{4}$ | $\frac{1}{3}$ $\frac{1}{2}$ $\frac{1}{4}$

3.

  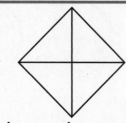

$\frac{1}{3}$ $\frac{1}{2}$ $\frac{1}{4}$ | $\frac{1}{3}$ $\frac{1}{2}$ $\frac{1}{4}$ | $\frac{1}{3}$ $\frac{1}{2}$ $\frac{1}{4}$

 Mixed Review

Write the number that is between.

4. 45, _____, 47   74, _____, 76   41, _____, 43

5. 57, _____, 59   69, _____, 71   21, _____, 23

6. 39, _____, 41   62, _____, 64   78, _____, 80

Problem Solving • Use Logical Reasoning

Cross out pictures that do not match the clues.
Circle the picture that matches the clues.
Use fraction circles to check.

1. There are 3 children. Each gets an equal share. How would you cut the pizza?

2. There are 2 children. Each gets an equal share. How would you cut the pizza?

3. There are 4 children. Each gets an equal share. Which shows one equal share?

4. There are 3 children. Each gets an equal share. Which pizza shows 3 equal parts?

5. There are 2 children. Each gets an equal share. Which shows one equal share?

Parts of Groups

Color to show each fraction.

1.
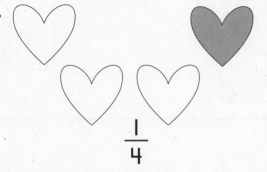
$\dfrac{1}{4}$

2.
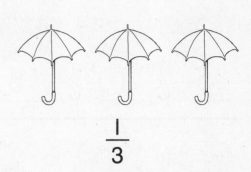
$\dfrac{1}{3}$

3.
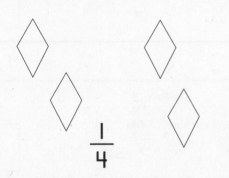
$\dfrac{1}{4}$

4.
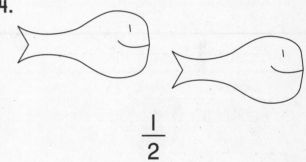
$\dfrac{1}{2}$

5.
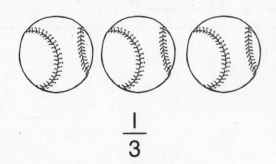
$\dfrac{1}{3}$

6.
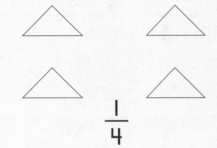
$\dfrac{1}{4}$

▶ **Mixed Review**

Write the sum or difference.

7. $3 + 4 =$ _____

10. $6 + 5 =$ _____

13. $4 + 6 =$ _____

8. $8 - 2 =$ _____

11. $9 - 4 =$ _____

14. $7 + 5 =$ _____

9. $12 - 6 =$ _____

12. $7 + 3 =$ _____

15. $11 - 4 =$ _____

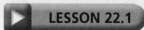

Pennies and Nickels

Count by ones or fives. Write the amount.

1.

_____ ¢, _____ ¢, _____ ¢, _____ ¢
5   10   15   20

20 ¢

2.

_____ ¢, _____ ¢, _____ ¢

☐ ¢

3.

_____ ¢, _____ ¢, _____ ¢, _____ ¢, _____ ¢

☐ ¢

4.

_____ ¢, _____ ¢, _____ ¢, _____ ¢, _____ ¢

☐ ¢

▶ **Mixed Review**

Write the numbers that are one less and one more.

5. _____ , 21 , _____

6. _____ , 59 , _____

7. _____ , 14 , _____

8. _____ , 40 , _____

Pennies and Dimes

Count by tens. Write the amount.

1.

___10___ ¢, ___20___ ¢

[20] ¢

2.

_____¢, _____¢, _____¢, _____¢

[ ] ¢

3.

_____¢, _____¢, _____¢, _____¢, _____¢, _____¢

[ ] ¢

4.

_____¢, _____¢, _____¢, _____¢, _____¢

[ ] ¢

▶ **Mixed Review**

Add or subtract.

5. $5 + 6 =$ ___ $3 + 6 =$ ___ $10 - 3 =$ ___

6. $3 + 9 =$ ___ $11 - 6 =$ ___ $7 + 4 =$ ___

Count Groups of Coins

Count. Write the amount.

1.   $\boxed{22}$ ¢

   _5_ ¢, _10_ ¢, _15_ ¢, _20_ ¢, _21_ ¢, _22_ ¢

2. $\square$ ¢

   ___ ¢, ___ ¢, ___ ¢, ___ ¢, ___ ¢, ___ ¢

3. $\square$ ¢

   ___ ¢, ___ ¢, ___ ¢, ___ ¢, ___ ¢, ___ ¢

4. $\square$ ¢

   ___ ¢, ___ ¢, ___ ¢, ___ ¢, ___ ¢, ___ ¢

▶ **Mixed Review**

Add.

5. $2 + 3 + 4 =$ ___

6. $3 + 2 + 3 =$ ___

7. $1 + 2 + 3 =$ ___

8. $4 + 2 + 3 =$ ___

Count Collections

Count. Write the amount.

1. ☐ ¢

2.
 ☐ ¢

3.

 ☐ ¢

4. ☐ ¢

▶ **Mixed Review**

Write the missing numbers.

5. 4 + ☐ = 7   6. 7 − 4 = ☐

7. ☐ + 4 = 9   8. 9 − 4 = ☐

9. 7 + ☐ = 11   10. 11 − 7 = ☐

Problem Solving • Make a List

Yoko wants to buy an apple for 25¢.

In what ways can she use ,

, and to make 25¢?

List five ways to make 25¢.

Use .

Draw and label the coins.

Ways to Make 25¢		
dimes	nickels	pennies
1. (10¢) (10¢)	(5¢)	0
2.		
3.	0	
4.		
5. 0		

Trade Pennies, Nickels, and Dimes

Use coins. Trade for nickels and dimes.
Use the fewest coins. Draw the coins.

1.

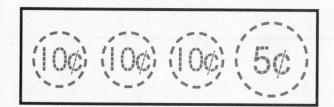

(10¢) (10¢) (10¢) (5¢)

2.

3.

▶ **Mixed Review**

Count by twos. Write the missing number.

4. 10, 12, 14, _____, 18, 20, _____, 24

5. _____, 6, 8, 10, 12, _____, 16, 18

6. 22, 24, 26, 28, _____, _____, 34, 36

Quarters

Count on from a quarter. Write the total amount.

1.

__25__ ¢, __30__ ¢, __35__ ¢

35 ¢

2.

_____ ¢, _____ ¢, _____ ¢, _____ ¢, _____ ¢

☐ ¢

3.

   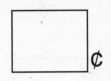

_____ ¢, _____ ¢, _____ ¢, _____ ¢

☐ ¢

▶ **Mixed Review**

Write the difference.

4. 8 – 5 = ____   7 – 4 = ____   6 – 4 = ____

5. 9 – 6 = ____   8 – 0 = ____   5 – 2 = ____

6. 7 – 5 = ____   6 – 3 = ____   4 – 3 = ____

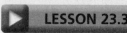

Half Dollar and Dollar

Draw and label the coins. Write how many.

1. Show how many nickels equal 1 dollar.

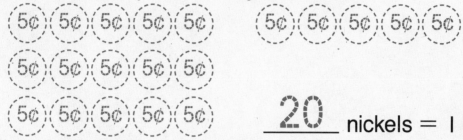

_____20_____ nickels = 1 dollar

2. Show how many dimes equal 1 dollar.

_____ dimes = 1 dollar

3. Show how many quarters equal 1 half dollar.

_____ quarters = 1 half dollar

▶ **Mixed Review**

Add or subtract.

4. $5 - 3 =$ _____ $7 - 3 =$ _____ $5 + 3 =$ _____

5. $7 + 3 =$ _____ $6 - 4 =$ _____ $5 - 4 =$ _____

6. $6 + 4 =$ _____ $5 + 4 =$ _____ $4 + 2 =$ _____

7. $8 + 2 =$ _____ $8 - 2 =$ _____ $4 - 2 =$ _____

Compare Values

Write the amount for each group.
Circle the amount that is greater.

1.

_____70_____ ¢   _____ ¢

2.

_____ ¢   _____ ¢

▶ **Mixed Review**

Add.

3. $4 + 5 =$ _____ | $10 + 5 =$ _____ | $8 + 9 =$ _____

4. $6 + 7 =$ _____ | $12 + 8 =$ _____ | $9 + 6 =$ _____

Same Amounts

Use coins.
Show the amount in two ways. Draw the coins.
Circle the way that uses fewer coins.

1.  | 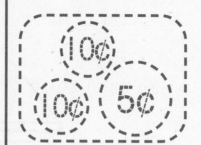 |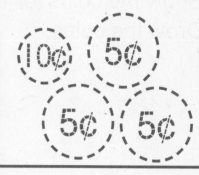

2.  | |

3. | |

► **Mixed Review**

Write the number that comes between.

4. 18, _____, 20   5. 29, _____, 31

6. 5, _____, 7   7. 44, _____, 46

8. 54, _____, 56   9. 36, _____, 38

Problem Solving • Act It Out

Show the coins for each amount.
Draw the coins.

1.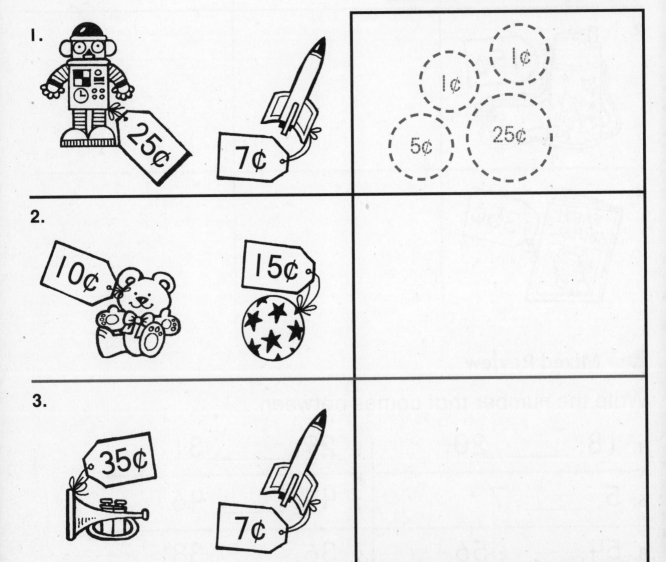

2.

3.

Read a Clock

Use a . Show each time.

Trace the hour hand. Write the time.

1. 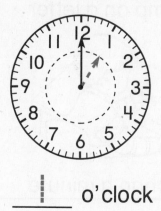	**2.** 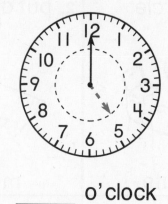	**3.** 
_____ o'clock	_____ o'clock	_____ o'clock
4. 	**5.** 	**6.** 
_____ o'clock	_____ o'clock	_____ o'clock

▶ **Mixed Review**

Count on from a quarter. Write the total amount.

7.

_____ ¢, _____ ¢, _____ ¢, _____ ¢ [ ] ¢

Problem Solving • Use Estimation

About how long would it take? Circle your estimate.
Then act it out to check.

1. put 10 chairs in a circle

(more than a minute)

less than a minute

2. put a stamp on a letter

more than a minute

less than a minute

3. open a door

more than a minute

less than a minute

4. write 10 spelling words

more than a minute

less than a minute

5. read a big book

more than a minute

less than a minute

6. sharpen a pencil

more than a minute

less than a minute

Time to the Hour

Use a . Show each time. Write the time.

1.

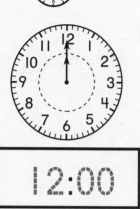

12:00

2.

:

3.

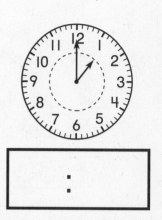

:

4.

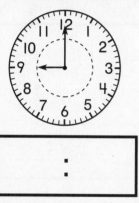

:

5.

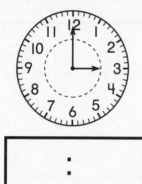

:

6.

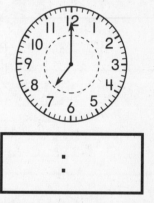

:

7.

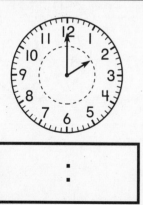

:

8.

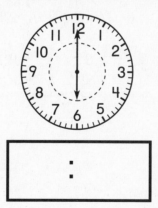

:

9.

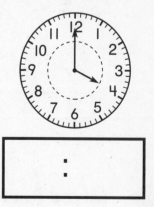

:

▶ **Mixed Review**

Write the sum.

10. $7 + 3 =$ ___ $4 + 3 =$ ___ $6 + 3 =$ ___

11. $0 + 3 =$ ___ $3 + 3 =$ ___ $2 + 3 =$ ___

Tell Time to the Half Hour

Use a 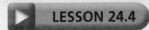 to show the time. Write the time.

1.

__5:30__

2.

___:___

3.

___:___

4.

___:___

5.

___:___

6.

___:___

7.

___:___

8.

___:___

9.

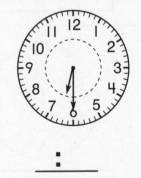

___:___

▶ **Mixed Review**

Circle the greater number.

10. 77 or 87	11. 23 or 13	12. 61 or 16
13. 45 or 54	14. 67 or 76	15. 39 or 99

Practice Time to the Hour and Half Hour

▶ **Vocabulary**

Draw the hour hand and the minute hand.

1.

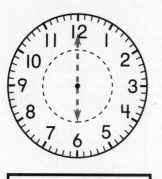

6:00

2.

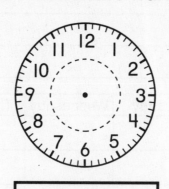

10:00

3.

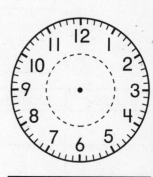

3:30

4.

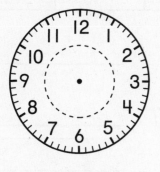

12:30

5.

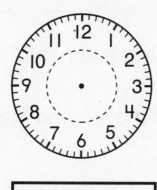

7:00

6.

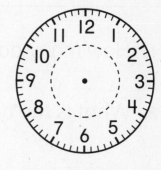

1:00

▶ **Mixed Review**

Which numbers come next? Count by fives or tens.

7. 30, 40, 50, _____, _____, _____

8. 15, 20, 25, _____, _____, _____

Use a Calendar

Fill in the calendar for next month.
Use the calendar to answer the questions.

_____

- - - - - - - - - - - - - - - - - - -

_____

Sunday	Monday	Tuesday	Wednesday	Thursday	Friday	Saturday

1. How many days are in the month?   _____

2. What is the date of the second Monday?   _____
                                            _____
                                            - - - - - - - -

3. What day of the week is the 20th?   _____

4. Color the Mondays | red ▷ .

5. Color the Fridays | yellow ▷ .

▶ **Mixed Review**

Subtract.

6. $12 - 8 =$ ___ $9 - 7 =$ ___ $10 - 3 =$ ___

7. $15 - 5 =$ ___ $7 - 2 =$ ___ $18 - 9 =$ ___

Daily Events

1. Today is Friday.
   Draw something special you
   could do during these times.

In the morning yesterday	In the afternoon today	In the evening tomorrow

▶ **Mixed Review**

Add.

2. $4 + 3 + 1 =$ ___ $5 + 2 + 4 =$ ___ $5 + 3 + 4 =$ ___

3. $2 + 3 + 4 =$ ___ $3 + 4 + 5 =$ ___ $4 + 3 + 3 =$ ___

Problem Solving • Make a Graph

Ask 10 classmates to choose the sport they like the best.

1. Make a tally mark for each choice. Then make a picture graph.

Sports We Like		Total
soccer	⚽	
softball	⚾	
football	🏈	

Sports We Like

	soccer ⚽	softball ⚾	football 🏈
10			
9			
8			
7			
6			
5			
4			
3			
2			
1			
0			

2. Which sport did the most classmates choose?

_____

_____

3. Which sport did the fewest classmates choose?

_____

_____

4. How many classmates chose softball or football?

_____ classmates

5. How many classmates chose soccer or softball?

_____ classmates

Read a Schedule

Saturday Activities	Start	End
clean room	(clock)	(clock)
hockey practice	(clock)	(clock)
lunch	(clock)	(clock)

Use the chart to answer the questions.

1. Which activity lasts the shortest amount of time?   lunch

2. Which activity lasts the longest amount of time?

3. Which activity is just before lunch?

▶ **Mixed Review**

Solve.

4. $4 + 4 =$ ___ $3 + 2 =$ ___ $5 + 6 =$ ___

5. $2 + 7 =$ ___ $1 + 6 =$ ___ $3 + 4 =$ ___

6. $9 + 2 =$ ___ $3 + 7 =$ ___ $3 + 5 =$ ___

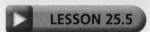

Name _____

Problem Solving • Make Reasonable Estimates

Circle the best estimate for each activity.

1.

brushing your teeth

about a week

about an hour

(about a minute)

2.

playing soccer

about a week

about an hour

about a minute

3.

jumping rope

about a week

about an hour

about a minute

4.

doing homework

about a week

about an hour

about a minute

5.

buying shoes

about a week

about an hour

about a minute

Compare Lengths

Put three in order from shortest to longest.
Draw them.

1. shortest

_____

2.

_____

3. longest

_____

▶ **Mixed Review**

Solve.

4. $2 + 2 =$ ___ $4 + 4 =$ ___ $6 + 6 =$ ___

5. $3 + 3 =$ ___ $1 + 1 =$ ___ $5 + 5 =$ ___

6. $4 + 2 =$ ___ $3 + 1 =$ ___ $5 + 3 =$ ___

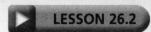

Use Nonstandard Units

Use real objects and ⬭.
Estimate. Then measure.
Circle the shortest object with | red |▷.
Circle the longest object with | blue |▷.

Object	Estimate	Measurement
1.	about _____ ⬭	about _____ ⬭
2.	about _____ ⬭	about _____ ⬭
3.	about _____ ⬭	about _____ ⬭
4.	about _____ ⬭	about _____ ⬭

 Mixed Review

Solve.

5. $6 + 5 =$ ___ $11 - 4 =$ ___ $5 + 3 =$ ___

6. $12 - 4 =$ ___ $8 + 2 =$ ___ $10 - 5 =$ ___

Inches

Use real objects and an inch ruler.
Estimate. Then measure.

Object	Estimate	Measurement
I.	about _____ inches	about _____ inches
2.	about _____ inches	about _____ inches
3.	about _____ inches	about _____ inches
4.	about _____ inches	about _____ inches

 Mixed Review

Solve.

5. $7 + 4 =$ ___ $5 + 4 =$ ___ $8 + 3 =$ ___

6. $10 - 4 =$ ___ $12 - 7 =$ ___ $11 - 5 =$ ___

Inches and Feet

About how long is the real object?
Circle the answer that makes sense.

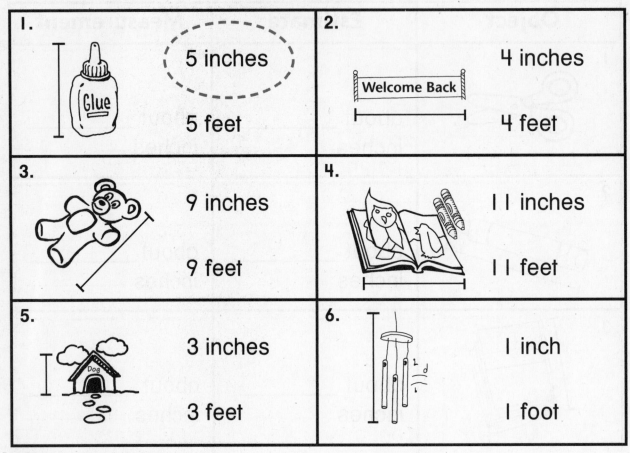

1. (5 inches)

    5 feet

2. 4 inches

    4 feet

3. 9 inches

    9 feet

4. 11 inches

    11 feet

5. 3 inches

    3 feet

6. 1 inch

    1 foot

 Mixed Review

Solve.

7.
$$\begin{array}{r} 5 \\ +3 \\ \hline \end{array} \qquad \begin{array}{r} 3 \\ +2 \\ \hline \end{array} \qquad \begin{array}{r} 4 \\ +2 \\ \hline \end{array} \qquad \begin{array}{r} 5 \\ +1 \\ \hline \end{array} \qquad \begin{array}{r} 3 \\ +2 \\ \hline \end{array}$$

8.
$$\begin{array}{r} 6 \\ +3 \\ \hline \end{array} \qquad \begin{array}{r} 4 \\ +1 \\ \hline \end{array} \qquad \begin{array}{r} 7 \\ +2 \\ \hline \end{array} \qquad \begin{array}{r} 1 \\ +3 \\ \hline \end{array} \qquad \begin{array}{r} 5 \\ +1 \\ \hline \end{array}$$

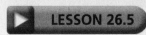

Centimeters

Use real objects and a centimeter ruler.
Estimate. Then measure.

Object	Estimate	Measurement
1.	about _____ centimeters	about _____ centimeters
2.	about _____ centimeters	about _____ centimeters
3.	about _____ centimeters	about _____ centimeters
4.	about _____ centimeters	about _____ centimeters

 Mixed Review

Solve.

5. $4 + 3 =$ _____ $6 + 2 =$ _____ $7 + 5 =$ _____

6. $12 - 8 =$ _____ $10 - 2 =$ _____ $11 - 6 =$ _____

Problem Solving • Make Reasonable Estimates

About how many beads long is the string?
Circle the answer that makes sense.

1.

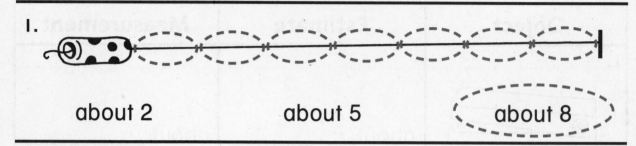

about 2   about 5   (about 8)

2.

about 2   about 4   about 6

3.

about 2   about 4   about 6

4.

about 3   about 5   about 8

▶ **Mixed Review**

Solve.

5. $5 + 4 =$ ___ $4 + 5 =$ ___ $6 + 3 =$ ___

6. $3 + 6 =$ ___ $8 + 3 =$ ___ $3 + 8 =$ ___

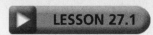
Use a Balance

About how many ⊂═⊃ does is take to balance?
Use real objects, a ⟁, and ⊂═⊃.
Estimate. Then measure.

Object	Estimate	Measurement
1. (quarter dollar coin)	about _____ ⊂═⊃	about _____ ⊂═⊃
2. (paintbrush)	about _____ ⊂═⊃	about _____ ⊂═⊃
3. (crayon)	about _____ ⊂═⊃	about _____ ⊂═⊃
4. (cube block)	about _____ ⊂═⊃	about _____ ⊂═⊃

5. Circle the heaviest object in blue.
6. Circle the lightest object in red.

> You used the smallest number of paper clips to balance the lightest object.

▶ **Mixed Review**

Solve.

7. $9 - 5 =$ _____ $3 + 2 =$ _____ $8 - 6 =$ _____

8. $5 + 6 =$ _____ $12 - 9 =$ _____ $7 + 5 =$ _____

Pounds

Estimate how much each object will weigh.
Then measure.

Object	Estimate	Measurement
1.	about ___ pounds	about ___ pounds
2.	about ___ pounds	about ___ pounds
3.	about ___ pounds	about ___ pounds
4.	about ___ pound	about ___ pound

5. Circle the heaviest item in blue.

6. Circle the lightest item in red.

▶ **Mixed Review**

Solve.

7. $9 + 0 =$ ___ $9 - 2 =$ ___ $11 - 3 =$ ___

8. $8 - 4 =$ ___ $10 + 2 =$ ___ $6 + 3 =$ ___

Name _____

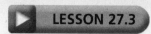

Kilograms

Estimate how much the real object will measure.
Use grams or kilograms. Then measure.

Object	Estimate	Measurement
1.	about_____ grams	about_____ grams
2.	about_____ kilogram	about_____ kilogram
3.	about_____ grams	about_____ grams
4.	about_____ grams	about_____ grams

▶ **Mixed Review**

Measure each object in inches and centimeters.

5.

about ____ inches   about ____ centimeters

6.

about ____ inches   about ____ centimeters

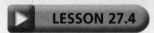

Problem Solving • Predict and Test

How many grams is the object?
Use ⬜ and ▭. Predict.
Then test.

1.

Predict: _____ grams

Test: _____ grams

2.

Predict: _____ grams

Test: _____ grams

3.

Predict: _____ grams

Test: _____ grams

4.

Predict: _____ grams

Test: _____ grams

5.

Predict: _____ grams

Test: _____ grams

Nonstandard Units

Use real containers and a ⬚.
Estimate. Then measure.

Container	Estimate	Measurement
1. SUNNY FARMS MILK	about ___ ⬚	about ___ ⬚
2. (cup)	about ___ ⬚	about ___ ⬚
3. Orange Juice	about ___ ⬚	about ___ ⬚
4. Cottage Cheese	about ___ ⬚	about ___ ⬚

▶ **Mixed Review**

Circle the number that is less.

5. 74 or 23   39 or 54   87 or 26

6. 44 or 35   84 or 12   60 or 81

7. 31 or 52   28 or 62   77 or 55

Cups, Pints, and Quarts

Estimate. Then measure.
Trace to name the size of the container.

Container	Estimate	Measurement	Size
1. ![orange juice carton]	about __ cups	about __ cups	_____ quart
2. ![takeout box]	about __ cups	about __ cups	_____ pint
3. ![milk carton]	about __ cups	about __ cups	_____ pint

 Mixed Review

Add.

4. $4 + 9 =$ ___ $5 + 6 =$ ___ $7 + 8 =$ ___

5. $6 + 6 =$ ___ $10 + 6 =$ ___ $12 + 6 =$ ___

Liters

Does each container hold less
or more than a liter?
Estimate. Then measure.

Think:
a liter

Container	Estimate	Measurement
1.	less than more than	less than more than
2.	less than more than	less than more than
3.	less than more than	less than more than
4.	less than more than	less than more than

▶ **Mixed Review**

Subtract.

5. $11 - 9 = $ ___ $12 - 3 = $ ___ $15 - 5 = $ ___

6. $14 - 7 = $ ___ $18 - 9 = $ ___ $17 - 8 = $ ___

7. $16 - 8 = $ ___ $15 - 9 = $ ___ $19 - 9 = $ ___

Temperature

Read the temperature.
Color the thermometer to show the temperature.

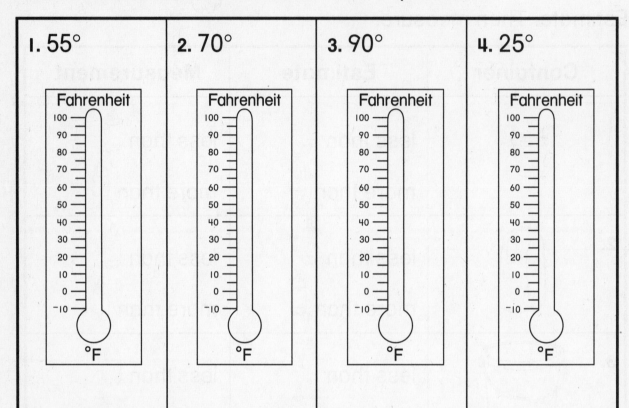

| 1. 55° | 2. 70° | 3. 90° | 4. 25° |

 Mixed Review

Count forward.

5. 67, _____, _____, _____, _____, _____, _____, _____, _____

6. 75, _____, _____, _____, _____, _____, _____, _____, _____

7. 16, _____, _____, _____, _____, _____, _____, _____, _____

Problem Solving: Choose the Measuring Tool

Circle the correct tool to measure.

1. Which apple is heavier?

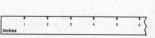

---

2. How tall are you?

 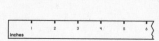

---

3. How wide is the book?

 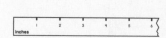

---

4. Which glass holds more?

---

5. How much does an
egg weigh?

 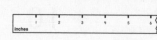

---

6. How much water will the
sink hold?

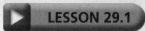
Use Mental Math to Add Tens

20 + 60 means
2 tens + 6 tens.

Think:

$$\begin{array}{r} 20 \\ + 60 \\ \hline 80 \end{array}$$

$$\begin{array}{r} 2 \text{ tens} \\ + 6 \text{ tens} \\ \hline 8 \text{ tens} \end{array}$$

Add.

1.

10	30	60	50	20
+ 50	+ 50	+ 30	+ 20	+ 30

2.

20	60	30	30	20
+ 40	+ 10	+ 30	+ 40	+ 20

3.

40	70	60	10	20
+ 50	+ 10	+ 20	+ 80	+ 70

▶ **Mixed Review**

Write >, <, or = in the circle.

4. $5 + 4 \bigcirc 6 + 3$ 5. $7 - 3 \bigcirc 6 - 1$

6. $2 + 2 \bigcirc 5 - 1$ 7. $8 - 4 \bigcirc 2 + 1$

Add Tens and Ones

Use Workmat 3 and [cube image] to add.
Write the sum.

1.

Tens	Ones
4	1
+	7
4	8

Tens	Ones

2.

Tens	Ones
2	5
+	3

Tens	Ones

3.

Tens	Ones
3	3
+	4

Tens	Ones

4.

Tens	Ones
2	7
+	1

Tens	Ones

▶ **Mixed Review**

Write the time.

5.

_____

6.

_____

7.

_____

8.

_____

Add Money

Add.

1. 17¢ + 12¢ ___ ¢	2. 25¢ + 24¢ ___ ¢	3. 40¢ + 12¢ ___ ¢	4. 35¢ + 11¢ ___ ¢	5. 23¢ + 35¢ ___ ¢
6. 45¢ + 3¢ ___ ¢	7. 55¢ + 14¢ ___ ¢	8. 53¢ + 35¢ ___ ¢	9. 62¢ + 5¢ ___ ¢	10. 31¢ + 28¢ ___ ¢
11. 30¢ + 43¢ ___ ¢	12. 25¢ + 23¢ ___ ¢	13. 14¢ + 25¢ ___ ¢	14. 42¢ + 35¢ ___ ¢	15. 28¢ + 30¢ ___ ¢
16. 72¢ + 6¢ ___ ¢	17. 43¢ + 12¢ ___ ¢	18. 52¢ + 16¢ ___ ¢	19. 34¢ + 20¢ ___ ¢	20. 33¢ + 33¢ ___ ¢

▶ **Mixed Review**

Write the amount.

21.

_____ ¢

22.

_____ ¢

Name _____

Use Mental Math to Subtract Tens

Think:

$$\begin{array}{r} 60 \\ -\ 40 \\ \hline 20 \end{array}$$

6 tens
− 4 tens
2 tens

60 − 40 means
6 tens − 4 tens.

Subtract.

1.
$$\begin{array}{r} 70 \\ -10 \end{array}$$
$$\begin{array}{r} 60 \\ -20 \end{array}$$
$$\begin{array}{r} 90 \\ -70 \end{array}$$
$$\begin{array}{r} 80 \\ -30 \end{array}$$
$$\begin{array}{r} 70 \\ -40 \end{array}$$

2.
$$\begin{array}{r} 50 \\ -20 \end{array}$$
$$\begin{array}{r} 90 \\ -30 \end{array}$$
$$\begin{array}{r} 50 \\ -10 \end{array}$$
$$\begin{array}{r} 20 \\ -20 \end{array}$$
$$\begin{array}{r} 80 \\ -10 \end{array}$$

3.
$$\begin{array}{r} 60 \\ -10 \end{array}$$
$$\begin{array}{r} 80 \\ -60 \end{array}$$
$$\begin{array}{r} 50 \\ -40 \end{array}$$
$$\begin{array}{r} 30 \\ -10 \end{array}$$
$$\begin{array}{r} 90 \\ -40 \end{array}$$

▶ **Mixed Review**

Write the sum.

4. $6 + 6 =$ _____ $2 + 2 =$ _____ $1 + 1 =$ _____

5. $9 + 9 =$ _____ $4 + 4 =$ _____ $8 + 8 =$ _____

6. $3 + 3 =$ _____ $7 + 7 =$ _____ $5 + 5 =$ _____

Name _____

Subtract Tens and Ones

Use Workmat 3 and  to subtract.
Write the difference.

Remember to subtract the ones first!

1.

Tens	Ones
2	8
−	5
2	3

Tens	Ones

2.

Tens	Ones
1	9
−	5

Tens	Ones

3.

Tens	Ones
3	5
−	2

Tens	Ones

4.

Tens	Ones
3	8
−	7

Tens	Ones

▶ **Mixed Review**

Write the sum.

5. $4 + 5 + 1 =$ ___ $3 + 7 + 1 =$ ___

6. $3 + 5 + 3 =$ ___ $6 + 1 + 4 =$ ___

7. $4 + 2 + 4 =$ ___ $2 + 4 + 2 =$ ___

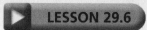

Subtract Money

Subtract.

1. 66¢ − 33¢ ____¢	2. 58¢ − 24¢ ____¢	3. 68¢ − 52¢ ____¢	4. 63¢ − 11¢ ____¢	5. 83¢ − 20¢ ____¢
6. 59¢ − 21¢ ____¢	7. 46¢ − 15¢ ____¢	8. 70¢ − 30¢ ____¢	9. 48¢ − 25¢ ____¢	10. 99¢ − 28¢ ____¢
11. 49¢ − 32¢ ____¢	12. 59¢ − 24¢ ____¢	13. 48¢ − 13¢ ____¢	14. 46¢ − 35¢ ____¢	15. 66¢ − 21¢ ____¢
16. 72¢ − 22¢ ____¢	17. 99¢ − 12¢ ____¢	18. 72¢ − 10¢ ____¢	19. 94¢ − 20¢ ____¢	20. 88¢ − 44¢ ____¢

▶ **Mixed Practice**

Write the numbers in order.

21. 75, 29, 44, 32, 65   ___, ___, ___, ___, ___

22. 83, 24, 64, 33, 10   ___, ___, ___, ___, ___

23. 33, 66, 99, 22, 55   ___, ___, ___, ___, ___

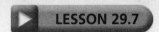

Problem Solving • Make Reasonable Estimates

Without adding or subtracting,
circle the closest estimate.

1. Chris buys an apple for
   36¢. He buys a banana
   for 29¢. About how much
   money does he spend in
   all?

   about 20¢

   about 70¢

   about 100¢

2. 45 children are in the gym.
   32 children leave. About
   how many children are still
   in the gym?

   about 10 children

   about 30 children

   about 50 children

3. Kate eats 7 nuts as a
   snack. She eats 6 nuts at
   lunch. She eats 4 nuts at
   dinner. About how many
   nuts does Kate eat in all?

   about 5 nuts

   about 15 nuts

   about 35 nuts

4. 18 spiders are on the tree.
   7 spiders crawl away.
   About how many spiders
   are left on the tree?

   about 10 spiders

   about 20 spiders

   about 30 spiders

Certain or Impossible

Draw an **X** to tell it is certain or impossible
to pull that shape from the bowl.

	Certain	Impossible
1. Pull a ⬤.		
2. Pull a ⬜.		
3. Pull a △.		
4. Pull a ▱.		
5. Pull a ⬜.		

Mixed Review

Add.

6. $9 + 5 + 3 =$ ___ $3 + 3 + 6 =$ ___

7. $5 + 4 + 3 =$ ___ $7 + 3 + 5 =$ ___

8. $2 + 3 + 4 =$ ___ $9 + 9 + 1 =$ ___

Name _____

More Likely, Less Likely

Draw a □ or a ○ to tell how likely each
shape is to be pulled from the bowl.

		More Likely	Less Likely
1.			
2.			
3.			
4.			
5.			

▶ **Mixed Review**

Solve.

6. $1 + 8 =$ ___ $9 - 8 =$ ___ | 7. $4 + 3 =$ ___ $7 - 3 =$ ___

8. $8 + 2 =$ ___ $10 - 2 =$ ___ | 9. $2 + 6 =$ ___ $8 - 6 =$ ___

Equally Likely

Draw a ◯, ▢, or △ to show which
shapes are equally likely to be pulled
from the bowl.

1.		
2.		
3.		
4.		
5.		

▶ **Mixed Review**

Subtract. Then write two related addition facts.

6. $17 - 8 =$ ___ | ___ + ___ = ___ | ___ + ___ = ___

7. $16 - 7 =$ ___ | ___ + ___ = ___ | ___ + ___ = ___

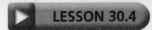

Problem Solving • Make a Prediction

Use a 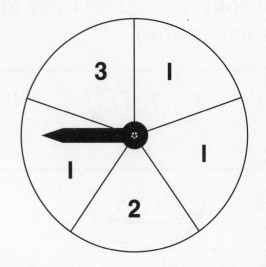 and a ⊂⊃
to make a spinner.

Predict. If you spin the
pointer 10 times, on which
number will it stop most often?
Circle that number.

   1   2   3

Check. Spin 10 times.
Make a tally mark after
each spin.
Write the totals.

	Tally Marks	Total
1		
2		
3		

1. On which number did your
pointer stop most often?

_____

2. Predict. If you spin
10 more times, on which
number will it stop most often?
Circle that number.

   1   2   3

Then spin to check.